Eco-Gardening

Eco-Gardening

Essential know-how and expert advice for gardening success

CONTENTS

CREATING AN ECO-GARDEN 006

PLANT FOR A BETTER WORLD 008

MAKING A SUSTAINABLE GARDEN 010

DESIGNING FOR WILDLIFE 012

HELPING TO HEAL THE PLANET 014

HOW TREES AFFECT THE CLIMATE 016

CHOOSING AND PLANTING A TREE 018

MAKING A MINI-WOODLAND 020

GREENING HARD SURFACES 022

CHOOSING PLANTS TO TRAP
POLLUTION 024

THE BENEFITS OF GARDEN PONDS 026

A WILDLIFE POND 028

HOW PLANTS PREVENT FLOODING 030

PLANTING FLOOD DEFENCES 032

GREAT TREES FOR SMALLER
GARDENS 034

ECO-GARDENING METHODS 038

RIGHT PLANT, RIGHT PLACE 040

NEW PLANTS FROM OLD:
SAVING SEEDS 042

EASY PROPAGATION METHODS 044

BUYING NEW PLANTS 046

PREPARING TO PLANT 048

CHOOSING COMPOSTS 050

MAKING YOUR OWN COMPOST 052

GROWING YOUR OWN VEGETABLES 054

PLANNING A VEGETABLE PATCH 056

GROWING YOUR OWN FRUIT 058

MAKING AN ECO-FRIENDLY LAWN 060

FEEDING PLANTS 062

KEEPING WEEDS UNDER CONTROL 064

KEEPING PESTS AT BAY 066

PEST DEFENCE 068

DEALING WITH DISEASES 070

PEST- AND DISEASE-RESISTANT PLANTS 072

REDUCE, REUSE, AND RECYCLE 078

REDUCING AND REUSING PLASTICS 080

MAKING YOUR OWN SEED POTS 082

ECO-FRIENDLY LANDSCAPE MATERIALS 084

CHOOSING AND MAKING FURNITURE 086

RECYCLED PLANTERS 088

RAISED BEDS FROM RECLAIMED
TIMBER 090

CHOOSING TOOLS AND EQUIPMENT 092

MAKING ECO-FRIENDLY EQUIPMENT 094

REDUCING YOUR GARDEN'S
FUEL CONSUMPTION 096

REDUCING WATER USE 098

COLLECTING AND REUSING WATER 100

MAKING A HOME FOR WILDLIFE 102

FEEDING GARDEN WILDLIFE 104

FOOD FOR BEES AND
OTHER POLLINATORS 106

FEEDING BUTTERFLIES AND MOTHS 108

PROVIDING FOOD FOR BIRDS 110

SOWING A MEADOW 112

INCREASING YOUR GARDEN'S HABITATS 114

PROVIDING WATER FOR WILDLIFE 116

CREATING A POND ECOSYSTEM 118

MAKING A HAVEN FOR BIRDS 120

HOMES FOR MINIBEASTS 122

MAKING INSECT HOTELS 124

PLANTS FOR POLLINATORS 126

PLANTS FOR BIRDS AND OTHER
GARDEN WILDLIFE 134

INDEX 140

BIBLIOGRAPHY AND RESOURCES 143

ACKNOWLEDGMENTS 144

Choosing plants such as alliums that attract bees and butterflies will help to boost the numbers of these important pollinating insects, which are threatened by loss of habitat and climate change.

CREATING AN ECO-GARDEN

Making a garden that is kind to the planet and to wildlife is easy when you know how. Planting it with trees, shrubs, flowers, and food crops will help to mitigate the effects of climate change, capture air pollutants, and create habitats for a vast number of creatures. Organic growing methods and a water source also keep wildlife thriving, while selecting furniture and equipment made from sustainable materials will help to lower your carbon footprint.

PLANT FOR A BETTER WORLD

While we feel instinctively that our gardens make us happier and healthier, recent scientific studies have proved that they offer these benefits and many others too. As well as boosting our mental and physical wellbeing, gardening can help to increase biodiversity by providing a home for a wide range of plants and wildlife. Filling our plots with trees, shrubs, and other plants also improves air quality and reduces pollution by mopping up the greenhouse gases that contribute to climate change. The good news is that just small changes to our gardens can help to make the planet a better place for everyone.

Deciduous trees and other plants absorb carbon dioxide, which contributes to climate change.

CLEANING THE AIR

Environmental organizations worldwide are campaigning for the planting of more woodlands and forests to help restore wildlife habitats and mitigate the effects of climate change by absorbing the greenhouse gas carbon dioxide (CO_2). If each of us were to plant just one tree in our garden, this would make a huge contribution to the fight against global warming. In fact, it may be more effective than the planting of a dense forest, which current research shows may not be as useful in reducing temperatures as smaller deciduous woodlands, particularly in cool and temperate regions. The benefits increase if we add shrubs and flowering plants, which in some cases can trap other air-borne pollutants too.

DIVERSITY MATTERS

The hundreds of thousands of plant species in our world each represent a tiny microcosm of life, nurturing many types of insects, birds, and other animals, including humans. As you would expect, trees and shrubs support more wildlife than smaller species, but all have a role to play. To maximize your contribution to the welfare of the creatures in your vicinity, include a broad spectrum to help nesting birds, amphibians, bees, and other pollinators, along with the minibeasts that nourish the soil beneath the surface. The distribution of wild plant species is also contracting due to urbanization, so growing a wide variety of native plants in the garden can help to safeguard their future too.

Planting a broad range of species will sustain many forms of wildlife.

There are plant species suited to nearly every spot in the garden, even areas that are heavily shaded by foliage or manmade structures.

FOOD FOR THOUGHT

Making space for fruit and vegetables to help feed your household will lower your carbon footprint by reducing food miles. Growing your own food is fun, and while producing something to eat year-round may be too ambitious in a small garden, you can still make a valuable contribution with a few pots of lettuces and strawberries or some cherry tomatoes in a hanging basket.

Cherry tomatoes in a basket make a beautiful edible feature.

FILLING THE GAPS

You do not need a large garden to accommodate a wide diversity of plants; pack them into every nook and cranny and you will be surprised by the number of wildlife visitors a small space can support. Let a patch of lawn grow to form a matrix of grasses and wild flowers that support a range of insect life. Plant roofs with tiny succulents and walls with ivy and other climbers that will insulate your home, reducing fuel costs as well as attracting wildlife. Fill the gaps between paving stones with plants rather than mortar and watch the insects move in, and use shade-lovers under benches or seats. As the old adage goes, nature abhors a vacuum, so squeeze in some life-supporting plants wherever you can.

HARDY VS TENDER PLANTS

The best way to ensure that your plants thrive is to choose those that like your site, soil, and local climate. This will result in healthy plants that are more resilient to attacks from pests and diseases and do not need chemical fertilizers to boost their performance. The hardiness of a plant is a measure of the lowest temperatures it will tolerate, and this is an important point to consider when selecting plants that will be outside all year. The different levels of hardiness are included in the plant profiles in this book.

FULLY HARDY This term is used to describe plants that will cope with temperatures below -15°C (5°F).

HARDY Some plants are not fully hardy, but are able to survive temperatures that fall below freezing; where this is the case, the lowest temperatures are stated, allowing you to check that your average winters will not be too cold for the particular species.

HALF-HARDY You may find this term used to describe some annuals and plants from warm areas. These can survive low temperatures and, in many cases, once mature they will tolerate 5°C (41°F) or lower, but they will not cope with frost.

TENDER This refers to tropical plants and those that do not thrive or will die in temperatures lower than 10–13°C (50–55°F).

Rudbeckia laciniata is a robust hardy plant that enjoys moist soil and sun.

MAKING A SUSTAINABLE GARDEN

The mantra for any eco-gardener is to reuse and recycle goods and materials wherever possible rather than buying them new. There are many imaginative ways in which to do this, from reusing old food containers for seedlings and plants to decorating your outdoor space with secondhand furniture or items you have made yourself. Reducing our reliance on plastic can be more challenging, but new biodegradable seed trays and plant pots are making it easier. Just pause before acquiring anything brand-new and consider other ways of making your garden beautiful and more sustainable.

SECOND CHANCES

When sourcing suitable containers and trays for your plants and seedlings, take a look in your kitchen cupboards for inspiration. Food cans, old biscuit tins, and plastic containers about to go into the recycling bin can all be repurposed for use in the garden. Giving these items a second life prevents them from ending up in landfill or being burned in polluting incinerators – sadly, more than 75 per cent of plastics that could be recycled are currently disposed of in this way. Also consider purchasing secondhand rather than new tools and furniture – wooden and metal pieces are good choices, and even if you have no proof that they were made from sustainable materials, reusing them prevents waste and reduces landfill.

Reused tables and chairs that are not a matching set look perfectly at home in an informal wildflower meadow.

A vintage pan can be repurposed to make a characterful plant container.

Biodegradable pots made from magazines and newspapers are perfect for vegetable seedlings and will quickly decompose in the soil.

REDUCING THE USE OF NON-BIODEGRADABLES

The gardening industry teems with products made from plastics and other non-biodegradable materials. Reusing plastic is more environmentally friendly than throwing it away and, depending on the type (see p.80), it may also be the only way of preventing it going to landfill or being burned. When buying new, look for products made from biodegradable materials such as wood, bamboo, wool, and seed husks. Many metals used in the garden do not biodegrade easily but they can be recycled an infinite number of times. Purchasing high-quality goods also reduces landfill because they do not need to be replaced frequently; spades, forks, and other tools can last a lifetime, or longer.

BE WISE WITH WATER

Capturing and reusing rainwater is a must for all eco-gardeners. It not only reduces the demand for tap water – which is treated with disinfectants and other chemicals and comes with a cost to the environment – but it is also better for plants and pondlife. Installing a water butt is the best way to capture the rain that pours off the roofs of your house, sheds, and other outdoor buildings; you will find information on installing these eco-friendly vessels and other ways of saving water on pp.98–101.

Water butts to store rainfall do not take up much space in the garden.

RECYCLING AND UPCYCLING

Reclaimed timber, bricks, and other building materials can be reused to make decking, walls, fences, and pergolas, and they will lend an aged patina to these features that blends beautifully into a garden setting. Reclamation and salvage companies

Recycled timber is used here to create a rustic path between plantings.

offer a wide selection of materials and you can also ask local craftspeople to fashion furniture and statuary from them to suit your needs. You may find suitable materials on local freecycle websites, too – but check that anything wooden you propose to purchase has not been treated with creosote or chromated copper arsenate (CCA), which are harmful to wildlife and the environment, and to our health too. In Reduce, Reuse, and Recycle (see pp.80–91), you will find more ways to give new life to old materials in your garden.

DESIGNING FOR WILDLIFE

Wildlife of all kinds is under threat from climate change, loss of habitat, and the use of pesticides, but we can support birds, insects, amphibians, and small mammals if we design our gardens with them in mind. Growing pollen- and nectar-rich flowers and fruiting trees and shrubs provides these creatures with food, while plants that offer hibernation and roosting sites protect them from predators and harsh weather.

Densely packed plants make good habitats for wildlife.

HOMES FOR INSECTS

Recent scientific studies show a worrying decline in insect populations worldwide. Apart from the collapse in the numbers of bees, there are many species on the critical or endangered list, with others becoming extinct. More than 95 per cent of all animals are insects, and we depend on them to pollinate our food crops, provide food for birds and other small animals, and recycle the dead plants and wildlife in the soil, thereby releasing the nutrients that plants need to grow. It is vital that we all do our bit to make good homes for these creatures in our gardens by planting pollen and nectar-rich flowers. Some insects also feed and breed on particular stems and leaves, so to produce an ideal habitat include a collection of trees, shrubs, and perennials, and leave areas aside for a few weeds to take root. For more advice on making insects welcome in your garden, see pp.122–133.

Plants that attract pollinators will help to boost their declining populations.

WATER WORLDS

One of the best ways to attract all sorts of wildlife is to include a water feature in your garden. Installing a pond will lure a wide range of species, from the tiniest beetles to bats and birds. Water also offers a home for plants and wildlife species that are adapted to live in this aquatic environment. While you will need to add the plants, frogs, toads, and other amphibians will appear as if by magic when you provide water, as will the aquatic insect life. Where space is tight, you can still offer this precious resource with a bird bath or shallow dish of water. Try the ideas on pp.28–29 and pp.116–119 for gardens large and small.

The catkins of silver birch are food for insects such as shield bugs.

A pond with a bog garden on its banks provides homes and food for amphibians and aquatic insects.

LIVING THE HIGH LIFE

There are few plants that help wildlife as much as trees. Providing nectar for pollinators in spring when they bloom, nesting sites later in the season when birds are looking for safe places, and fruit and leaves for a variety of minibeasts and birds to eat, they are hard to beat. Large shrubs and hedges provide similar benefits and when planted together, they create a fantastic range of habitats. You will find advice on planting trees on pp.18–19 and a selection to choose from for smaller gardens on pp.34–37.

WILD ABOUT NATURE

Introducing benefits for wildlife often means doing less in the garden. Nothing could be easier than leaving a pile of autumn leaves and tree and shrub prunings in a quiet corner behind a shed, which will provide the perfect home for many hibernating garden creatures and minibeasts that feast on decaying wood. A patch of weeds such as nettles can draw in breeding butterflies and moths, and leaving fallen fruits on the ground will provide a valuable food source for them in late summer and autumn. Refraining from using pesticides and herbicides will make your garden richer too, not only by increasing insect life but also potentially reducing air pollution. Scientists have found that particles from pesticides used in Asia were blown to the west coast of the US in less than a week, and while gardeners may be using them on a much smaller scale, the cumulative effect could be just as widespread and harmful.

Piles of prunings offer a home for minibeasts and hibernating amphibians.

Dense planting of varied species, from trees to ground-covering and aquatic plants, helps to absorb carbon dioxide from the atmosphere and also provides homes and food for wildlife.

HELPING TO HEAL THE PLANET

Every gardener can play a part in reducing pollution levels. The plants we grow absorb carbon dioxide, one of the greenhouse gases responsible for global warming. Shrubs and trees can also trap the tiny particulates that are a significant component of air pollution, especially in built-up areas such as towns and cities, and all plants help to prevent flooding, which can result in toxins entering rivers and oceans. This chapter shows how you can use your garden to maximize these benefits and help to heal the planet.

HOW TREES AFFECT THE CLIMATE

Trees can help to protect the health of our planet in many ways. They absorb carbon dioxide (CO_2), the greenhouse gas produced by fossil fuels that has led to a rise in global temperatures, while also releasing oxygen into the atmosphere. Trees regulate localized temperatures, too, helping to keep our homes warmer in winter and cooler in summer, thereby reducing heating and air-conditioning needs.

Planting trees in your garden will help to reduce air pollution.

Hawthorns (_Crataegus_) help to mitigate the effects of climate change, while also offering food and shelter for insects and birds.

HOW TREES CAPTURE CARBON

Trees and other plants absorb carbon dioxide (CO_2) to make food through a process known as photosynthesis. During this process, plant cells convert the carbon from CO_2 into sugars, which they store in their leaves, roots, and stems and use as fuel for growth. At the same time, as a by-product of photosynthesis, plants release oxygen back into the atmosphere, thereby creating the carbon cycle that all life depends upon.

Due to their size, trees absorb relatively more CO_2 than other plants, which is why they play such an important role in the fight against climate change. The destruction of forests worldwide and our dependence on fossil fuels have contributed to an excess of carbon, which acts like a greenhouse, trapping heat and raising global temperatures. While recent research shows that the picture is complex, since trees absorb sunlight and emit a range of chemicals that have both cooling and warming effects on the atmosphere, experts agree that, on balance, planting trees helps to lower CO_2 levels and reduces the effects of this pollutant.

LOCAL TEMPERATURE REGULATORS

As well as absorbing carbon, trees help to reduce temperature levels indirectly through their ability to moderate localized conditions. Their canopies cast shade, which has a marked effect on temperature. Trees also draw up large volumes of water through their roots and release most of it as water vapour through their leaves and stems, a process known as evapotranspiration, which further cools the air.

Town planners are starting to recognize the effect that trees can have on air quality. Studies of Atlanta in the US show that the paved-over city centre is 5–8°F (2.6–4.3°C) hotter than the tree-lined suburbs. This is known as the urban heat island effect. The higher temperatures are caused by a lack of vegetation, waste heat from homes, shops, and offices, and dark surfaces such as tarmac on roads absorbing sunlight.

Conversely, planting deciduous trees in the path of prevailing winds can help to keep buildings warmer in winter. The

TOP TIP PLANT A DECIDUOUS TREE TO SHADE A SOUTH-FACING WINDOW; THE CANOPY WILL COOL YOUR HOME DURING THE SUMMER WHILE THE BARE STEMS WILL ALLOW SUN THROUGH TO WARM THE PROPERTY IN WINTER, THUS REDUCING THE NEED FOR BOTH AIR CONDITIONING AND HEATING.

Black walnut (*Juglans nigra*) is a large tree with handsome autumn foliage.

leafless canopies allow sunlight through to windows, while the network of stems gives the building protection from biting winds. By cooling homes in summer and insulating them in winter, trees help to reduce energy needs and pollution from heating and air conditioning.

The trees lining boulevards in Paris help to lower streetside temperatures.

TREES AS A FOOD SOURCE

Growing fruit and nut trees can help to reduce CO_2 levels on a small scale by providing a food source that has almost no carbon footprint. Perfect for a small garden, a fruit tree will take up very little space if you buy a dwarf type, yet can provide many benefits – some even offer more than one type of fruit, such as apples and pears, on a single plant. These trees also create excellent wildlife habitats, providing food for pollinators in spring and for birds and butterflies in autumn. (For advice on choosing a tree, see *pp.18–19*.)

An apple tree can provide an abundant harvest for both people and wildlife.

CHOOSING AND PLANTING A TREE

There is a tree to suit every garden, whether you have a tiny courtyard, an average-sized plot, or several acres that would accommodate a small woodland. Consider what type of tree you would like and then check your site, soil, and the space you have available to ensure your choice will thrive in your garden. Following the planting advice here will help to get your tree off to a good start.

Sessile oak (*Quercus petraea*) acts as a host for more than 250 species of insect, which offer a food source for birds.

SELECTING A TREE

The first step is to match your site and soil (*see pp.40–41*) with a tree that will be happy in your garden. Also check the final height and spread to make sure there is enough space for it to grow. Fruit trees are grafted on to different rootstocks that determine their size, so you should be able to find one suitable for your space – but note that many trees are not self-fertile, which means that you will require two in order to guarantee a crop of fruit.

If you have only a small garden, an acer will grow happily in a large pot.

CHOOSING THE SITE

All trees, however small, will cast shade. Consider where and when the sun hits your garden (*see p.41*) and place a ladder or other tall structure where you plan to site the tree in order to see exactly where its shade will fall throughout the day. Planting one to partially shade a patio can be ideal, but a large tree that plunges an area into semi-darkness for most of the day will limit your other plant choices. Also, beware of planting close to your property, as the root system of

Himalayan cherry (*Prunus rufa*) has peeling bark and white blossom.

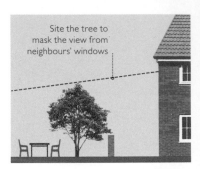

Site the tree to mask the view from neighbours' windows

A well-placed tree can be used both to cast some shade on a seating area and to provide privacy from neighbours.

most trees is usually about equal to the width of the canopy. A small tree that will grow 3m (10ft) wide can be safely planted 4–5m (12–15ft) from the house.

Leave a gap between your tree and your boundary, too, as the soil close to walls and fences tends to be in a rain shadow and very dry, which will hinder the plant's growth. The canopy will also spread over your neighbour's garden, which may cause problems as it matures. Ideally, plant the tree at least half its final width from the boundary. Also make sure that your tree does not block the sun from your neighbour's garden. Planting it closer to your seating areas may actually provide more privacy from their windows (see diagram above).

PLANTING A TREE

1 Leave the tree in its pot in a bucket of water for about an hour to soak the rootball. Meanwhile, dig a hole three times as wide as the pot and the same depth. Use a fork to loosen the soil around the sides of the hole.

2 Place the tree in the hole and, using a cane laid across the top, check that the point where the roots meet the stem will be level with, or slightly proud of, the soil surface.

3 Remove the tree from its pot and use your fingers to gently loosen the roots coiled around the side of the rootball. Place it in the hole.

4 Fill in with soil around the rootball and use your toe to gently press it down to remove any air pockets. Water well. Apply a 5–7.5cm (2–3in) layer of organic material, such as bark chips or well-rotted compost, as a mulch over the rootball, leaving a 10cm (4in) gap around the stem.

5 Large trees will need staking. Hammer in a sturdy stake at a 45° angle on the side of the tree opposite to the prevailing wind direction. Attach with a tree tie.

6 Water trees regularly during dry spells for 3–5 years after planting to ensure good root growth. Drench the area once or twice a week so that water reaches the roots at lower levels rather than giving small quantities more often, as this can encourage roots to grow towards the surface where the soil is drier.

NEED TO KNOW
- The best time to plant a tree is in mid- to late autumn.
- Do not plant if the ground is waterlogged or frozen.
- Loosen the stake ties as the tree grows.
- Remove the stake after 2–3 years when the roots are established.

MAKING A MINI-WOODLAND

The plants that grow beneath a tree create a valuable layer of vegetation which plays an equally important role in reducing air pollution and CO₂ levels. The foliage and flowers also offer a habitat for wildlife and **pollinators. For inspiration on naturalistic planting, visit a deciduous woodland at different times of the year to see how the rich tapestry of plants growing there can be used in your own garden.**

BENEFITS OF WOODLAND PLANTING

All land-based plants absorb carbon from the atmosphere in the form of carbon dioxide (CO_2) when they photosynthesize to make food (*see p.16*). While individual trees may take up more than smaller plants, adding a carpet of other species beneath them will increase the benefits. In fact, some studies show that these plants may contribute as much or even more than trees in offsetting climate change in areas other than the tropical zones around the equator.

The understorey planting beneath a tree also performs a valuable role in the creation of a sustainable woodland ecosystem. The plants' roots and leaves help to prevent soil erosion, which in turn protects the tree from damage during storms and periods of drought. They also absorb the nutrients released from decomposed leaves that fall from the tree canopy, which means that once established they will not need extra fertilizers to thrive. Both the tree and the plants beneath it create an insulating blanket that nurtures wildlife, such as insects, worms and other soil-borne decomposers, and small garden creatures.

Dead nettle, hardy geraniums, and barrronwort thrive in dappled shade.

PLANTING IN LAYERS

Look to nature for inspiration on what to plant beneath trees. The floor of deciduous woodland comprises layers of plants that make use of the variable light. Spring-flowering shrubs such as viburnums, mahonias, and sweet box (*Sarcococca confusa*), together with bulbs including snowdrops, bluebells, and wild garlic, bloom when light levels are highest, before foliage blocks the sun. A much longer list of plants that offer habitats and food for wildlife grow later in the dappled light at the edge of the tree canopy when the foliage has unfurled (*see opposite*). Try filling the space beneath your tree with shrubs interplanted with smaller woodland bulbs, biennials, and perennials.

Here a colourful carpet of shade-lovers includes hairy chervil (*Chaerophyllum*), red campion, and ferns.

WOODLAND PLANTS FOR GARDENS

SHRUBS

Daphne (*Daphne* species) • Spindle (*Euonymus alatus* and *E. europaeus*) • Mahonia (*Mahonia* species) • Sweet box (*Sarcococca confusa*) • Guelder rose (*Viburnum opulus*) • Weigela (*Weigela* species)

SPRING BULBS

Wild garlic (*Allium ursinum*) • Wood anemones (*Anemone nemorosa*) • Dog's tooth violets (*Erythronium*) • Snowdrops (*Galanthus*) • English bluebell (*Hyacinthoides non-scripta*) • Siberian squill (*Scilla siberica*)

BIENNIALS AND PERENNIALS

Bugle (*Ajuga reptans*) • Cow parsley (*Anthriscus sylvestris*) • Hart's tongue fern (*Asplenium scolopendrium*) • Bergenias • Foxgloves (*Digitalis purpurea*) • Male fern (*Dryopteris filix-mas* and *D. affinis*) • Barronwort (*Epimedium*) • Wood spurge (*Euphorbia amygdaloides*) • Sweet woodruff (*Galium odoratum*) • Dusky cranesbill (*Geranium phaeum*) • Hellebores (*Helleborus*) • Dead nettle (*Lamium*) • Soft shield fern (*Polystichum setiferum*) • Primrose (*Primula*) • Red campion (*Silene dioica*) • Foam flower (*Tiarella cordifolia*)

Snowdrops such as *Galanthus* 'S. Arnott' bloom before trees are in full leaf.

HOW TO PLANT WOODLANDERS

Plan your planting carefully before you start. Include plants that will grow in dry shade, such as male ferns, sweet box, and wood spurge, closest to the tree, but leave a space of 1–1.2m (3–4ft) between them and the trunk. Spring bulbs can also be planted at about this distance from the tree – they may self-seed closer to it if conditions are favourable. Use plants that prefer dappled shade and damper conditions at the edge of the tree canopy.

Before planting, check each plant's spread and measure out the distances between them so that they all have space to reach their full potential. Plant at the same depth the plants were at in their pots, and water all of them well for a few months after planting while the roots are establishing. Some plants, such as sweet woodruff and dead nettle, spread via long roots and may need to be kept in check by removing excessive growth in spring.

Space plants to allow them to grow to their full potential beneath the canopy.

MAKING THE MOST OF YOUR SPACE

Squeeze as many plants as you can into your mini woodland. Where congested tree roots do not allow planting into the ground, add a pot of shade-loving, pollinator-friendly container plants such as single-flowered fuchsias or heliotropes.

***Fuchsia* 'Checkerboard'** is a deciduous shrub with single flowers.

GREENING HARD SURFACES

Make the most of the ecological benefits that plants offer by squeezing them into every nook and cranny in your garden. Green roofs and walls add extra leafy layers that enrich a wildlife haven, while ribbons of flowers and foliage growing between paving stones also help to offset your carbon footprint. Additional planting such as this will increase food sources and habitats for bees, birds, and other garden wildlife.

LEAFY COVER-UPS

As well as planting up traditional beds and borders, use other surfaces to squeeze in more foliage and flowers that will improve the air quality and reduce pollution levels. Cover house walls and boundary fences with pollinator-friendly climbers such as honeysuckle (*Lonicera periclymenum*) and wisteria. Ideal for small gardens, these climbers take up very little ground space yet offer many benefits. As well as absorbing carbon (*see p.16*), when grown against a house wall they will help to keep your home cool in summer and offer some insulation in winter, thereby lowering fuel bills. Their tall, twining stems and leaf cover provide birds with nesting and roosting sites while their pollen attracts beneficial insects.

A variety of colours and leaf shapes makes a green wall visually appealing.

MIND THE GAPS

The cracks between the stones or a groove chiselled into the top of a wall add yet more opportunities for planting. Ferns may self-seed naturally in walls located in cool, dark areas, or you can encourage them to colonize these structures by removing some pointing or a stone, wrapping a young plant in clay-rich soil and gently inserting it into the gap. In sunny areas, try houseleeks (*Sempervivum*), sedums, and other succulents in a wall.

Houseleeks and sedums planted into a wall make an eco-friendly feature.

GREEN WALLS

There are many ways in which to create a beautiful vertical garden on a wall or fence. One of the easiest methods is to fix specially designed planting pockets to your chosen structure and then simply fill them with compost and plants. However, while green walls can offer a good opportunity to increase the biodiversity in your garden, they do require frequent watering and feeding to thrive, so before installing one you should ensure you have an eco-friendly water supply and sufficient time to keep your wall in good health. Small, drought-tolerant plants, such as sedums, sea thrift (*Armeria maritima*), pot marigolds (*Calendula officinalis*), and heucheras are good choices for green walls.

The flowers of *Rosa* 'Rambling Rector' are followed by hips that birds feed upon.

PLANTING YOUR PAVING

Most gardens need some functional hard surfaces, particularly for dining and seating areas and paths that receive the most wear and tear. However, even these can be transformed into small-scale wildlife havens. Low-growing, spreading plants, such as thyme, Mexican fleabane (*Erigeron karvinskianus*), and the trailing bellflower (*Campanula poscharskyana*), will soon form a carpet between paving stones. Space the pavers in a path or patio (which will also make them more floodproof; see p.32), and fill the gaps with free-draining soil. Add your plants and top with a layer of gravel.

Plant low-growing species between patio slabs to increase wildlife value.

Perfect for paths, creeping thymes bounce back if stepped on occasionally.

Make a mini-meadow on a roof by constructing a shallow box on it, then fill with free-draining infertile soil, grasses, and wild flowers.

UP ON THE ROOF

Make the most of the roof space on your shed, garage, garden office, and even your home by planting sedums or wildflowers and grasses. Ensure your roof is strong enough to take the weight of the soil and plants before you start, particularly if you are planning to plant up your house roof. Most specialist suppliers of green-roof plants will be able to advise you. Sedum roofs are the easiest option and, for a shed or outbuilding, you can buy matting that you simply tack onto the structure. For larger plants, such as grasses and wildflowers, you will need to construct a shallow box on the roof that can hold a greater depth of compost and has good drainage to prevent waterlogging.

PLANTING CHOICES FOR SMALL GREEN ROOFS

New Zealand bur (*Acaena affinis*) • Sea thrift (*Armeria maritima*) • Oregano (*Origanum* species) • Sedums (*Sedum* species) • Houseleeks (*Sempervivum* species) • Creeping thyme (*Thymus serpyllum*)

NEED TO KNOW

- Check that the load-bearing capacity of your roof is sufficient for the soil and plants.
- Ensure that the roof is at a slight angle so that water drains into a gutter with a downpipe and water butt.
- Use lightweight, organic compost designed for roofs or a free-draining sandy soil.
- Choose plants that are non-invasive and low-growing.
- Always ask for expert advice if you are considering the installation of a green roof on your house rather than a garden building.

CHOOSING PLANTS TO TRAP POLLUTION

Few of us can escape the effects of the airborne pollution from vehicle exhausts and industry that permeates our towns, cities, and roadsides. However, shrubs and trees can help to protect us by trapping the tiny pollution particles on their leaves. Some plants are more effective than others, so if you live in an urban area or close to a busy road, include some of the best to create a pollution-busting barrier.

A dense hedge screens pollution and makes a natural backdrop to a border.

PLANTS THAT FIGHT AIR POLLUTION

Pollution caused by diesel and petrol vehicle emissions and the burning of fossil fuels is linked to a range of illnesses, including asthma, cancer, and heart disease. The good news is that the leaves of shrubs and trees can trap tiny particles from these emissions (particulates), thereby improving the local air quality and helping to keep us healthy. If you do not have enough space for a hedge or screen, a fence covered in a climber such as ivy (*Hedera helix*) will also offer good protection.

The ridged and grooved leaves of a hornbeam hedge trap particulates, helping to keep a garden free from pollution.

A tall yew hedge is highly efficient at trapping particulates.

LEAVES THAT TRAP PARTICULATES

Research by the Royal Horticultural Society has found that some shrubs and small trees trap particulates more effectively than others. For example, a yew (*Taxus baccata*) hedge can trap four times as many particulates as a photinia screen because its waxy leaves are more effective at capturing and embedding pollutants than smooth foliage. Plants with hairy leaves, such as flowering currants (*Ribes*), are other good choices for a pollution barrier because their hairs create a greater surface area and trap the particulates between them. Studies also show that conifers with scaly leaves are efficient at trapping particulates, while the small ridges and grooves of rough-leaved plants, such as hornbeam (*Carpinus betulus*) and hawthorn (*Crataegus monogyna*), do likewise.

POLLUTION-ABSORBING PLANTS

Japanese barberry (*Berberis thunbergii*) • Birch (*Betula*) • Hornbeam (*Carpinus betulus*) • Hawthorn (*Crataegus monogyna*) • Cypress (*Cupressus*) • Spindle (*Euonymus japonicus*) • Ivy (*Hedera helix*) • Holly (*Ilex aquifolium*) • Bird cherry (*Prunus pagoda*) • Firethorn (*Pyracantha*) • Flowering currant (*Ribes sanguineum*) • Elder (*Sambucus*) • Yew (*Taxus baccata*) • Red cedar (*Thuja plicata*) • Viburnum (*Viburnum tinus*)

Hawthorn (*Crataegus monogyna*) is effective at screening particulates.

The hairy leaves of flowering currant (*Ribes*) trap pollutants.

PLANTING A HEDGE

A permeable hedge that allows air to flow through it is the most effective feature for trapping particulates. To create one in your garden, choose plants from the list above that suit your site and soil conditions. Plant them at about 90cm (3ft) intervals in a prepared bed, free of weeds and large stones or debris. Prune immediately after planting and then annually in autumn or late winter, before birds begin to build their nests. Cut only the sides until the plants reach their desired height – research shows that a screen of about 2m (6ft) is ideal. When your screen is the right height, trim the top as well to keep it to that size.

While evergreens such as yew and holly provide a year-round barrier to pollution, they can eventually create a dense screen that allows little air to pass through. The best solution to trapping pollutants may be to plant a variety of shrubs and trees: this has the added benefits of creating multiple habitats and increasing biodiversity.

NEED TO KNOW

- Check that your chosen shrubs suit your site and soil.
- Space plants at 90cm (3ft) intervals in a prepared bed.
- Plant at the same depth as the plants were in their original pots. If you are planting bare-root shrubs, look for the dark soil line just above the rootball.
- Choose some plants that also offer wildlife benefits, such as pollen-rich flowers and berries for birds.

Plant hedges 40–45cm (16–18in) from a wall or fence.

THE BENEFITS OF GARDEN PONDS

Ponds draw many forms of wildlife into the garden to drink, bathe, and reproduce in the fresh water, while the plants that thrive in these aquatic environments add yet another layer of biodiversity. Some experts say that ponds can help to trap carbon and reduce the levels of this element and other greenhouse gases in the atmosphere, although research into these effects is still ongoing.

PONDS AS THRIVING ECOSYSTEMS

A pond provides a huge boost to your garden's biodiversity. Not only does it broaden the range of plants you can grow, it can attract and support a large number of local wildlife species, including frogs, toads, and other amphibians. Birds and small garden creatures will also be drawn to the water, and damselfly and dragonfly species may visit your pond too, especially if it is fringed with plants.

In addition to these creatures, ponds are also home to a host of smaller aquatic species that live at or below the surface, including water snails, water beetles, and pond skaters, all of which contribute to a healthy ecosystem. In fact, two-thirds of all freshwater species are supported by garden ponds.

A variety of plants bordering the pond adds to the range of wildlife it attracts.

Providing areas of clean water may attract dragonflies to the garden.

Birds will visit a pond to drink and bathe, and some also eat aquatic insects.

PONDS AS CARBON SINKS

Some experts have found that ponds may play a role in mitigating climate change by acting as carbon sinks. One study suggests that they may be better at storing carbon than either woodland or grassland, despite covering a proportionally much smaller area. The scientific research demonstrates that ponds are not only much richer in species compared to rivers, streams, and lakes, they also bury carbon in the litter that accumulates at the bottom. Carbon dioxide (CO_2) produced by microbes as they break down autumn leaves and dead pond flora and fauna is locked in the sediment rather than being released into the atmosphere, where it would contribute to global warming. As with all plants, aquatic species also use carbon dioxide during photosynthesis (see p.16), which helps to lower levels of this greenhouse gas.

A pond edged with naturalistic planting blends easily into the garden and may have considerable environmental benefits.

In Canada, research into farm ponds found that the water was acting as a sink for the greenhouse gas nitrous oxide, produced when nitrates from fertilizers are washed into the water. However, studies showing that ponds help to reduce air pollution are not conclusive; other research has found that lakes and some ponds actually contribute to greenhouse gas emissions. Nevertheless, ponds definitely benefit wildlife and are well worth including in your garden.

Partly covering the pond surface with plants helps to maintain oxygen levels.

MAINTAINING A HEALTHY POND

Ponds are sensitive ecosystems and need careful planning to ensure they remain healthy. Any excess nutrients, such as nitrogen from fertilizers, can cause a chemical imbalance that leads to excessive algae growth and poor oxygen levels. To avoid these problems, do not use fertilizers, including organic matter such as garden compost, close to the pond or where rain may wash the nutrients into it. Whenever possible, use rainwater from a butt to top up levels rather than water from the tap, which contains chlorine.

Pond water also needs a rich supply of oxygen to sustain the wildlife in it. Introducing aquatic plants such as waterlilies as well as marginals that are adapted to the shallows around the edges (see pp.44–47) helps to regulate oxygen levels. The amount of sunlight also plays a part because warm water contains less oxygen, so make sure the pond surface is partly shaded in order to keep it cooler.

In winter, ice and snow can act as a blanket, keeping the deeper water warm enough for most pond life to survive. However, the reduced light reaching submerged plants can prevent them from photosynthesizing (see p.16) and releasing oxygen, which may kill overwintering creatures such as frogs, so sweep away snow to allow in more sunlight. Making a hole in the ice by placing a pan of hot water on the surface at the edge of the pond will also provide garden creatures with drinking water.

TOP TIP THE TEMPERATURE AND OXYGEN LEVELS WILL REMAIN MORE CONSISTENT IN A LARGE POND WITH A DEEP AREA IN THE MIDDLE, SO MAKE YOURS AS BIG AS POSSIBLE TO MAXIMIZE ITS WILDLIFE POTENTIAL. MOVING WATER, SUCH AS A SMALL CASCADE OR FOUNTAIN, WILL ALSO INTRODUCE MORE OXYGEN INTO THE WATER.

WILDLIFE PONDS

Providing the perfect habitat for a whole host of birds, insects, and small animals, a pond also adds a beautiful ornamental feature, with colourful plants and reflected light creating a dramatic focal point. A pond is easy to make over a weekend, and can be large or small, depending on the size of your plot. Sloping or stepped margins allow easy access for creatures to come and go, while the plants offer cover from predators. Fill your pond with rainwater from a butt (see *p.100*) or, if you use tap water, leave it for a couple of days before adding plants to allow chemicals such as chlorine to evaporate.

Planted pond margins create a perfect habitat for aquatic creatures, such as frogs and toads, as well as a host of insects.

SITING YOUR POND

Choose a sheltered location in an area where you can enjoy your pond from a seating area or your windows. Check that it is well away from utility pipes and cables, and not too close to overhanging trees or large shrubs. These would not only cast too much shade over the surface, restricting your choice of pond plants, but their leaves could also pollute the water if they fell into it in large quantities in autumn. Avoid making a pond near yew trees; their leaves will poison pond wildlife.

Aquatic bags filled with garden soil free from fertilizers and planted with marginals fit neatly onto pond shelves.

Choose an open site away from tall trees and shrubs that cast deep shade.

CREATING A HAVEN FOR WILDLIFE

Creatures of all kinds are drawn to water features, and a pond that has been designed especially for wildlife will greatly increase the biodiversity in your garden. Make a haven for small mammals, amphibians, and birds by creating a beached edge at one end of the pond, or use plants in pots to create shallow steps around the sides that allow them to get in and out of the water safely.

A fringe of pond plants in the shallows and along the margins will provide a leafy canopy that protects wildlife from the eyes of predators. Creating different water levels also allows you to include the widest range of plants, such as water lilies and other aquatics, and pollen-rich flowering marginals (see *pp.118-119*). These will help to keep the water clear, sustain pollinators, and make a beautiful year-round feature for you to enjoy. To give your pond a head start, in the spring ask friends if they have any frog spawn you can add to it.

MAKING A POND

Wildlife ponds do not need to be huge, but a deep area that remains frost-free offers creatures a safer home in winter.

YOU WILL NEED Hosepipe • Spade • Long length of wood • Builder's spirit level • Old carpet or proprietary pond underlay • Butyl pond liner • Sharp knife

1 Mark out the shape of your pond with a hosepipe. The bigger the pond, the more wildlife you will be able to home.
2 Dig out the pond to a depth of 45cm (18in), with sloping sides. Leave a shelf about 30–45cm (12–18in) wide at this depth around the edge. Dig out a central area 1m (3ft) deep and an adjacent area about 75cm (30in) deep.
3 Place a spirit level on a straight plank of wood to check that the pond will be level. Repeat in a few places around the pond edge, building up or removing soil as necessary.
4 Remove large or sharp stones in the walls and base. Line the pond with a proprietary underlay or an old carpet that has not been treated with chemical cleaners.
5 Cover the underlay with the butyl liner (see below), centring it over the hole. Push it down in the middle, and pleat along the sides and base to fit.
6 Fill the pond with water – ideally rainwater stored in water butts. Trim the liner with a sharp knife, leaving 45cm (18in) excess around the edges, which you can disguise with soil and plants, turf or rocks, taking care not to tear the liner.

LINER QUANTITIES Calculate the size of the butyl liner you will need as follows: measure the pond at its longest length (L) and its widest width (W); then the depth at its deepest point (D). The area of the liner needed is given by (2D+45cm+L) x (2D+45cm+W).

HOW PLANTS PREVENT FLOODING

Climate change has influenced global weather patterns, causing more extreme events such as summer storms and floods. Using the power of plants, gardeners can help to reduce flooding, which has wide-reaching effects on the environment and can lead to polluted rivers and oceans. Replacing paving with permeable surfaces, including lawns and gravel, can also help in the fight against flood damage.

Using permeable surfaces rather than paving for paths in your garden allows water to drain away easily.

Pollutants lead to algae which depletes the water of oxygen when it decomposes.

CAUSES AND EFFECTS

As the climate has warmed, extreme weather events such as storms and prolonged rainfall have increased. These bring both widescale and localized flooding, which can devastate ecosystems: plants and wildlife die in waterlogged conditions, and raw sewage can enter rivers and oceans when drains overflow in areas where rainwater and domestic sewage systems are not separated. Floodwater may also carry animal waste, engine oil, fertilizers, pesticides, and detergents. The effects of flooding are exacerbated in urban areas with little plant cover, where stormwater runs off hard surfaces into overstretched sewers. However, gardeners can make a difference by designing their gardens with flood prevention in mind.

WHY FLOODS OCCUR AFTER A STORM

The rate at which rain falls as well as the quantity has an effect on the flood risk. When water hits a surface more quickly than the ground can absorb it, such as during a storm, the rainwater runs off and flooding can occur, whereas gentle rain allows the soil more time to soak it up. After prolonged rainfall, the soil can also become saturated and if more rain then falls, the excess water will immediately produce surface runoff and flooding. By taking up water from the soil, plants continually prevent the soil from reaching saturation point, whereas hard, impermeable surfaces that have very little capacity to absorb rainwater will flood much more easily.

Where gardens lack sufficient permeable surfaces, water from heavy rainfall gushes out on to public roads.

PLAN PERMEABLE SURFACES

Integrate permeable surfaces into your garden wherever possible, rather than using paving stones or concrete – the latter is definitely not recommended, as it has one of the highest carbon footprints of all landscaping materials. Grass, aggregates such as gravel, and shredded prunings allow rainwater to penetrate to the soil, and these can also be used between pavers where a hard surface is needed. In some countries, you must by law have permeable surfaces in a garden, especially at the front of a property, where replacing gardens with hardstanding for cars has contributed to urban flooding. Check with your local planning department for guidelines.

TOP TIP ADD A LAYER OF LEAFMOULD OR ORGANIC HOMEMADE COMPOST, BOTH OF WHICH ARE HIGHLY POROUS, OVER YOUR BEDS AND BORDERS. NOT ONLY WILL THEY HELP TO INCREASE THE DRAINAGE CAPACITY OF THE SOIL, THEY ALSO PREVENT IT FROM FORMING A HARD IMPERMEABLE LAYER THAT RAIN WILL QUICKLY RUN OFF.

Wooden planks and gravel create an informal permeable pathway through a naturalistic scheme of ornamental grasses and low-growing herbs and perennials.

PLANTS TO THE RESCUE

Trees and other plants play a significant role in flood prevention. Tree canopies trap rainwater on the leaves and stems and slow the rate at which it falls to the ground. The plants beneath further reduce the volume of water that reaches the ground, and leaf litter on the soil surface holds water droplets for longer still. When rainwater finally reaches the ground, the soil soaks it up like a sponge, preventing it from running off the surface and causing flooding. Once the water is in the soil, plants draw it up through their roots and into their stems and leaves, where the excess is lost through tiny pores in the foliage called stomata – a process known as transpiration. The roots also create channels in the soil that allow rain to filter down into it more quickly.

Maximizing the planting area of your garden, particularly in areas where storm water may collect and run into the road and sewage systems, can significantly alleviate the risk of flooding (for ideas on how to incorporate more planting, see pp.32–33).

Trees and plants prevent flooding by absorbing rainwater before it can run off.

Rainwater is trapped by the leaves and branches

The tree draws up water from the roots and releases it back into the atmosphere

Plants beneath the tree trap more water as it falls from the canopy

Roots open up channels for water to pass through

PLANTING FLOOD DEFENCES

There are many ways to maximize the planting in your garden to reduce the risk of localized flooding and prevent the wider environmental damage it causes. Assess the areas that face the road, where stormwater will run off into the sewers, to see if there is scope for more plants. Climbers on walls and fences and groundcover plants squeezed between pavers offer opportunities in small gardens, while trees and large shrubs will absorb even greater volumes of water if you have space for them. Try some of the simple solutions here to create a leafy defence that will protect against water run-off.

PLANTED PATHWAYS

When you are planning a new pathway, include some planting in your design. Grass paths are absorbent, but they can rapidly deteriorate into muddy tracks on a route that is used every day, such as one leading to a house door. The soil beneath the grass will soon become compacted in these areas, killing the turf and creating an impermeable surface for water to run off. Where foot traffic is highest, combine hard landscaping materials such as wooden planks or stone pavers with gravel or another absorbent material. Space the planks or pavers at a comfortable stepping distance from one another and fill in between with the gravel, into which you can plant low-growing or spreading plants including thyme, Mexican fleabane (*Erigeron karvinskianus*), chamomile (*Chamaemelum nobile*), or houseleeks (*Sempervivum*). Fringe your path with larger plants to make a water-absorbent blanket that will prevent run-off.

A parking space can combine hard surfacing with areas of planting.

PARKING SOLUTIONS

Front gardens often double as parking spaces, but even these can include ribbons of planting between the hard areas needed for car tyres. Install tracks made from wood, bricks, or paving, and fill the areas in between with gravel and plants. Low-growing plants that will tolerate the occasional pressure of a car tyre running over them include creeping jenny (*Lysimachia nummularia*), bugle (*Ajuga reptans*), and thymes such as *Thymus serpyllum*. A sprinkling of forget-me-nots will add some colour in spring and self-seed into the gravel. Water the plants frequently after planting until they are well-established.

A combination of wood and gravel allows water to filter into the ground.

In areas with low foot traffic, grass paths make good water-absorbent walkways.

POTS FOR PAVED AREAS

Where hard materials are necessary, such as a doorstep or dining area, use plants in pots to mop up excess rainwater. The larger the pot the better, as these will hold more compost and absorb the most water. Small shrubs such as *Skimmia japonica* and *Euonymus fortunei*, ivies, and colourful violas are idea for shady doorsteps, or try lavender, sedums, and sedges (*Carex*) for sunnier spots. These will offer a decorative display and only require watering a couple of times a week during dry weather if they are planted in a large container such as a half-barrel.

Plants in pots of different sizes look attractive and will soak up water.

COLLECTING STORM WATER

In gardens where rainwater runs off patios or other hard surfaces and pools at the lower end of your plot, consider installing a mini rain garden. This is a depression filled with resilient plants where water can collect during a storm, rather than running off into the road or waterways. The planted area holds the excess water, while the roots create channels that allow it to drain slowly into the soil.

Dig a wide, saucer-shaped hole about 45cm (18in) deep, or 60cm (2ft) deep in clay soils that take longer to drain. Create a berm (lip) about 30cm (12in) wide and 10cm (4in) high, with some of the excavated soil along the lower edge of the hole to help capture water as it flows down the slope. Partly refill the hole so that it forms a slight depression, using a mix of homemade compost, excavated soil, and horticultural gravel, which will aid drainage. Add plants that are able to tolerate periods of waterlogging as well as drier conditions. Seek professional help if your garden is on a steep slope of more than about 12 per cent or 1:8.

This mini rain garden is filled with tough plants and topped with stones.

PLANTS FOR A MINI RAIN GARDEN

Bugle (*Ajuga reptans*) • False spirea (*Astilbe* species) • Sedges (*Carex* species) • Wood spurge (*Euphorbia amygdaloides*) • Hardy geranium (*Geranium* species) • Hostas (*Hosta* species) • Japanese water iris (*Iris ensata*) • Eulalia (*Miscanthus sinensis*) • Knotweed (*Persicaria affinis*) • Black-eyed Susan (*Rudbeckia* species)

GREAT TREES FOR SMALLER GARDENS

With so many trees to choose from, this catalogue of compact species and varieties is designed to help you select the right one for a small or medium-sized garden. Compact trees are ideal for beginners as they are less likely than large species to outgrow their allotted space or need expert pruning – yet they offer the same benefits of trapping and storing carbon dioxide, mitigating the effects of climate change.

SMOOTH SERVICEBERRY *AMELANCHIER LAEVIS*

HEIGHT AND SPREAD up to 8 × 8m (25 × 25ft)
SOIL Moist but well-drained
HARDINESS Fully hardy
SUN ☼ ☼

Grow this small deciduous tree for its coppery-pink spring leaves, which turn green in summer before taking on yellow and red autumn tints. The fragrant, starry, white spring blossom is followed in summer by edible, sweet, blue-black fruits, which are loved by birds. It makes a beautiful feature tree amid informal planting in a small city garden, or in a meadow area in a larger space. Undemanding, it will grow in most soils and needs very little aftercare once it is established.

Starry white blossoms are followed by blue-black fruits that attract birds to the garden.

FIELD MAPLE *ACER CAMPESTRE*

HEIGHT AND SPREAD up to 12 × 4m (40 × 12ft)
SOIL Well-drained/moist but well-drained
HARDY Fully hardy
SUN ☼ ☼

Often used as a hedge plant, this deciduous tree produces five-lobed, red-tinted leaves in spring which turn mid-green towards summer. They then take on yellow and red hues in autumn. The tiny greenish-yellow flowers attract polllinators in spring, and small mammals eat the winged seedheads in autumn. Tolerant of a wide range of sites and soils, this maple makes a beautiful addition to a wildlife garden. While it can grow tall, pruning the main trunk when the tree is young will limit its growth.

This maple's handsome five-lobed foliage turns yellow and red in autumn.

RIVER BIRCH *BETULA NIGRA*

HEIGHT AND SPREAD up to 12 × 7m (40 × 22ft)
SOIL Well-drained/moist but well-drained
HARDINESS Fully hardy
SUN ☼ ☼

Also known as the red or black birch, this impressive tree has shaggy pinkish-brown and white bark that makes an eye-catching feature in winter and green, diamond-shaped leaves that turn buttery-yellow in autumn. Its yellow spring catkins add to its charms. Ideal for boggy or wet soils, this tree is equally happy in drier conditions, and provides a good habitat for insects and birds. Buy a multi-stemmed tree, which will limit its growth, and plant it in a border or close to a natural water feature.

Buttery-yellow autumn leaves and colourful peeling bark are prized aspects of the river birch.

JUDAS TREE *CERCIS SILIQUASTRUM*

HEIGHT AND SPREAD 7 × 7m (22 × 22ft)
SOIL Well-drained
HARDINESS Survives all but the harshest winters
SUN ☼ ☀

This deciduous small tree is popular for its rose-lilac, pea-like flowers, which appear in late spring. These are followed by bright green, heart-shaped leaves and dark purple-tinted, flattened seedpods in midsummer. The leaves then turn gold in autumn. As well as attracting bees and other pollinators, the tree is able to feed itself by extracting nitrogen from the air via bacteria on its roots – nitrogen is essential for healthy leaves. It makes a beautiful feature in small courtyard gardens.

Pea-like flowers appear on the bare branches in spring, creating an eye-catching feature.

CORKSCREW HAZEL *CORYLUS AVELLANA* 'CONTORTA'

HEIGHT AND SPREAD 5 × 5m (15 × 15ft)
SOIL Well-drained
HARDY Fully hardy
SUN ☼ ☀

The unusual twisted branches of the corkscrew hazel make a striking focal point in winter and early spring, when the golden catkins appear and dangle from the branches like glittering earrings. The green leaves are slightly twisted too and turn yellow in autumn, when the tree produces nuts (although not as many as a standard hazel tree). Moth caterpillars feed on the leaves, while many types of bird eat the nuts. Underplant it with spring bulbs and pollen-rich summer-flowering perennials to add more wildlife benefits.

The twisted branches of this hazel give the impression of an ancient gnarled tree.

HAWTHORN *CRATAEGUS LAEVIGATA*

HEIGHT AND SPREAD up to 7 × 7m (22 × 22ft)
SOIL Well-drained/moist but well-drained
HARDINESS Fully hardy
SUN ☼ ☀

Supporting a wide range of insects and birds, this deciduous tree also offers a range of ornamental features. Clusters of creamy-white or pink flowers, loved by pollinators, appear after the green lobed leaves have unfurled. Then, in autumn, birds feast on the bright red berries, and the grooved bark and twisted stems add interest in winter. This tough little tree looks at home in an informal planting scheme and among wild flowers. It is also perfect for an exposed site or coastal garden.

Pink double flowers appear on thorny stems on the popular hawthorn cultivar 'Paul's Scarlet'.

SPINDLE TREE *EUONYMUS EUROPAEUS*

HEIGHT AND SPREAD 3 × 2.5m (10 × 8ft)
SOIL Well-drained
HARDY Fully hardy
SUN ☼

A good choice for a large pot in a courtyard, this deciduous tree produces dark green leaves and tiny pollen-rich flowers. While it is quite unassuming for much of the year, in autumn its colours trump almost everything else in the garden. The leaves fire up to a blazing scarlet, while orange-pink, winged fruit add to the spectacle and remain on the tree after the foliage has fallen. Plant it where its autumn guise can be enjoyed, with spring and summer flowers offering interest at other times.

Stunning autumn colour creates a dazzling feature in a small eco garden.

COMMON HOLLY *ILEX AQUIFOLIUM*

HEIGHT AND SPREAD up to 12 × 4m (40 × 12ft)
SOIL Well-drained/moist but well-drained
HARDY Fully hardy
SUN ☼ ☀

Characterized by its spiny evergreen foliage, holly offers interest all year round, with cultivars bearing variegated leaves providing added colour. The spring flowers attract pollinators, while the dense canopy makes an ideal nesting site for birds. Plant holly in a mixed border of shrubs and flowering perennials and annuals. For berries, which provide birds with food, choose a female form and ensure there is a male nearby to fertilize the flowers, or select a self-fertile variety such as 'J. C. van Tol'.

'Argentea Marginata' is a female form with cream-edged leaves, tinged pink in spring.

APPLE TREE *MALUS DOMESTICA*

HEIGHT AND SPREAD up to 4 × 4m (12 × 12ft)
SOIL Well-drained/moist but well-drained
HARDINESS Fully hardy
SUN ☼

Few trees beat the eating apple for great value in a small garden. As the green leaves unfurl in spring, pink or white pollen-rich blossom attracts honey bees; in autumn, the ripe apples are ready for picking, offering you a rich store of fresh fruit. The fallen apples also provide a feast for butterflies, birds, and other garden creatures. Buy tiny trees on dwarfing rootstocks for a patio pot or larger specimens to make a feature in a lawn or border. Late-flowering varieties are best in frost-prone areas.

Fresh apples taste delicious and offer a wonderful resource for garden wildlife.

JAPANESE CRAB APPLE *MALUS × FLORIBUNDA*

HEIGHT AND SPREAD up to 7 × 7m (22 × 22ft)
SOIL Well-drained/moist but well-drained
HARDY Fully hardy
SUN ☼ ☀

This pretty crab apple bears abundant pearly-pink flowers, opening from mid- to late spring just as the small, bright green leaves appear. The blossom is followed in autumn by small golden-yellow fruits that often remain on the tree into winter; the fallen apples offer a valuable food source for a range of wildlife. Tolerant of air pollution, this spreading tree makes a beautiful focal point in an informal garden underplanted with spring bulbs and summer perennials, or in a meadow of wild flowers and grasses.

A profusion of red buds open into pink flowers in spring, appearing with the new leaves.

MEDLAR *MESPILUS GERMANICA*

HEIGHT AND SPREAD up to 6 × 8m (20 × 25ft)
SOIL Well-drained/moist but well-drained
HARDINESS Fully hardy
SUN ☼ ☀

This spreading tree is perfect for eco-gardeners looking for a distinctive focal point. The medlar has large white spring flowers that are magnets for bees and the leathery leaves are joined by brown, apple-shaped fruits in autumn. These are hard and bitter, but if harvested in late autumn and left to ripen, they become softer and sweeter. The fallen fruits offer a winter food source for wildlife, too. Due to their spreading habit, medlars are best grown as standard or half-standard trees with a clear trunk.

Golden autumn leaves are complemented by the medlar's brown, apple-shaped fruit.

FUJI CHERRY *PRUNUS INCISA*

HEIGHT AND SPREAD 2.5 × 2.5m (8 × 8ft)
SOIL Moist but well-drained/well-drained
HARDY Fully hardy
SUN ☀

This compact flowering cherry is ideal for a small urban garden. A profusion of crimson buds opens to reveal clear white flowers, which are covered with bees and other pollinators when they appear in spring. The narrow, pale green leaves are tinged with bronze as they unfurl and also put on a show in autumn, when they turn bright orange and red before falling. The small fruits attract birds. Plant it where you can admire the foliage, as a specimen or in a mixed informal bed. Prune in early summer if necessary.

The profuse white flowers make this tree popular with various pollinating insects.

PEAR TREE 'CONFERENCE' *PYRUS COMMUNIS*

HEIGHT AND SPREAD 3 × 2.5m (10 × 8ft)
SOIL Well-drained
HARDY Fully hardy
SUN ☀

Pear trees offer a multitude of benefits for people and wildlife, including beautiful white spring blossom and juicy, edible fruits in autumn. For a small, manageable tree that bears Conference pears, choose one grown on a Quince 'C' rootstock; for a heavy crop, plant another pear tree nearby. The flowers lure bees and other pollinators, while butterflies and birds, including thrushes and blackbirds, will gorge on the fallen fruits in autumn. Plant among wild flowers in a meadow or in a mixed border.

The sweet fruits that make pear trees popular also help to feed garden wildlife.

ROWAN *SORBUS AUCUPARIA*

HEIGHT AND SPREAD 12 × 8m (40 × 25ft)
SOIL Well-drained/moist but well-drained
HARDINESS Fully hardy
SUN ☀ ☀

Also known as the mountain ash, this small to medium-sized tree offers many benefits for both gardeners and wildlife. Its flat clusters of small white flowers attract pollinators in late spring and the orange-red autumn berries provide an excellent source of food for birds. The green, divided leaves also take on fiery hues in autumn. In a small space, choose a multi-stemmed tree or compact cultivar such as 'Vilmorinii' and use your tree to add seasonal highlights to an informal planting scheme.

Bright red berries appear in large clusters in autumn and provide a rich feast for birds.

COMMON LILAC *SYRINGA VULGARIS*

HEIGHT AND SPREAD 7 × 7m (22 × 22ft)
SOIL Well-drained/moist but well-drained
HARDY Fully hardy
SUN ☀ ☀

A deciduous small tree or large shrub, lilac is celebrated for its large, conical heads of highly scented spring flowers that are a magnet for bees and other pollinators. There is a huge range of cultivars with flowers in shades of dark red, pink, purple, and white. The mid-green, heart-shaped foliage adds to the tree's charms. Plant it at the back of an informal mixed border or along a boundary with other shrubs and trees to add privacy to your garden. To keep it in shape, prune hard in early summer.

A lilac tree in full bloom is a stunning sight, while the scent of the flowers is an added bonus.

Harvesting seeds from crops such as peas and beans rather than buying them each year reduces your carbon footprint and the cost of the plants you grow.

ECO-GARDENING
METHODS

The way you grow your ornamental plants and crops has an impact on the environment. Using organic methods that do not rely on chemical fertilizers or the use of pesticides and herbicides helps to create healthier soil and ensures fewer pollutants wash into our waterways. Growing plants from seed, using biodegradable pots, and making compost at home also helps to reduce our carbon footprint. Explore these and other eco-friendly ways to raise and care for your plants.

RIGHT PLANT, RIGHT PLACE

The best way to ensure your plants thrive and to minimize the effects of pests and diseases is to grow those that enjoy your garden environment. This means finding out what type of soil you have and how much sun your plot receives through the day at different times of the year, then planting species adapted to those conditions. Most plant labels and catalogues give enough information to help you make the right choices.

Helenium autumnale **'Helena Red Shades'** needs a sunny spot to thrive.

CHECKING YOUR SOIL TYPE

Most garden soils fall into one of two categories: those rich in sand and those where clay is the main ingredient. Sand particles are relatively large and water is able to drain quickly through the spaces between them. This is why sandy soils are free-draining and usually quite infertile, as plant nutrients are soluble and dissolve in rainwater. By comparison, clay particles are tiny and trap moisture in the minute gaps between them. These soils are dense and prone to waterlogging, but tend to be more fertile. The best garden soil is loam, which contains almost equal measures of sand and clay. It retains enough water for plant roots to absorb, while allowing excess moisture to drain away – a condition referred to as "moist but well-drained". However, whatever your soil type, there will be plants to suit it.

To test your soil, dig up a small sample from just below the surface, leave it to dry off a little if wet, then roll it between your fingers.

SANDY SOIL When rolled between the fingers, sandy soil feels gritty and falls apart when you try to mould it into a ball or sausage shape. It is also generally pale in colour.

CLAY SOIL Smooth and dense, clay soil retains its shape when it is moulded into a ball or sausage shape. Soil that has a very high clay content will not crack even when it is bent into a horse-shoe shape.

Clay soil is dense and retains its shape when you manipulate it, which means you can mould it into sausage shapes or balls.

Dig up a little soil and roll it between your fingers to discover the consistency.

TESTING SOIL ACIDITY

The acidity or alkalinity of a soil is known as its pH value. This is important because some plants, such as rhododendrons and camellias, will only grow well in acid soils, while others, including lavender and rock roses, prefer alkaline conditions. Many plants are not fussy and catalogues often specify pH needs only when a plant is particularly demanding. You may also find acid-loving plants listed as lime-hating or lime-intolerant. Kits for testing soil pH are easy to use and it is well worth spending time assessing the soil in a few areas of the garden to check whether it is acid, alkaline, or neutral.

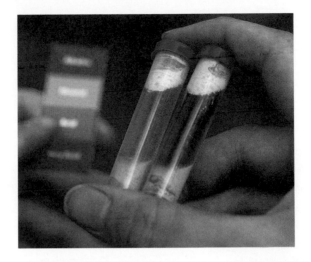

To use a soil testing kit, put the required amount of soil in the container, add the chemical powder and solution, and shake to mix the contents. When the soil has settled, compare the colour of the liquid against the chart.

CHECKING FOR SUN AND SHADE

It is vital to check where the sun falls on your garden throughout the day, as plants sited in the wrong light conditions may grow poorly, fail to flower, or, in the worst cases, die. When they are grown in shade, plants that enjoy bright sun, such as those with big, blowsy flowers, will become tall and leggy as their stems stretch towards the light. Plants that prefer shade, which include those with large, dark green leaves, will scorch if grown in too much light.

To find out which way your garden faces and, therefore, how much sun it receives, stand with your back to the house and use a compass. If you are facing south, the garden will be sunny for most of the day in summer; if the compass points north, the garden will be shady. Those facing east or west will receive the most sun in the morning or evening respectively. Remember that trees and buildings will cast shade in an otherwise sunny plot.

Another way to assess where the sun falls in the garden is to take photographs at different times of the day and during different seasons. Mark the areas of light and shade on a simple overhead plan.

In a garden centre, plants will have labels with information about their needs.

READING PLANT LABELS

Once you have completed an assessment of the conditions in your garden, look through catalogues and nursery websites for plants that will be happy and healthy there. Many websites offer easy-to-use search engines that will throw up lists of plants for different sites and soils. At a garden centre, check the labels on plants. These will specify the plant's needs, as well as its final height and spread and, often, if it is good for pollinators. Once you are sure that a plant is suitable, check that it looks healthy and is free of pests and diseases.

N

Morning

In the morning the west boundary will be in sun

In the evening the west boundary will be in shade

Noon

Evening

N

Morning

Noon

Evening

A north-facing garden will have shaded areas for much of the day, so locate trees where they will not rob you of sun.

In a south-facing garden the area directly in front of the rear boundary will be in some shade for most of the day.

NEW PLANTS FROM OLD: SAVING SEEDS

Harvesting seeds from existing plants to make new ones is a great way to lower your carbon footprint and save money, too. The seeds of many flowers, fruit, and vegetables are easy to harvest and will germinate and grow without too much effort – after all, that is what they are naturally designed to do. Some may not look exactly the same as their parents, but experimenting to see what may come up is all part of the fun.

COLLECTING SEEDS

Many beautiful plants and tasty crops will grow easily from seed harvested from their pods. Just remember that plants grown from seeds harvested from F1 hybrids, which have been bred to produce a specific flower or fruit colour, for example, may not look exactly like the parent plant. If you are not sure about yours and no longer have the seed packet, check online.

FLOWERS Once they are pollinated, flowers produce seedheads, the capsules that protect the seeds while they ripen before opening to release them. When the seedheads have changed from green to black, brown, or red, remove the whole pod or tap the plant gently above a paper bag – the seeds will fall out easily when they are ripe. Tip out the seeds onto a clean, dry surface and remove any bits of seed pod or other debris (chaff), which may harbour pests or diseases. Either sow the ripe seed soon after collecting it in autumn or store it and sow in spring.

FRUITS AND BEANS For soft fruits and fruiting vegetables, such as tomatoes, remove the fruit from the plant, scrape or scoop out the seeds and place them in a kitchen sieve. Clean the seeds by washing off the flesh, then leave them

Hollyhock (*Alcea rosea*) has large seeds that are easy to extract from the seedheads once they have turned brown.

to dry in a warm, dry place. Pea and bean pods can be left on the plant to ripen, or removed when dry and placed in a warm area to continue ripening.

STORING SEEDS Place dry seeds in clean paper bags or envelopes and label them with the plant name and date. If they are from tender plants, store them in a dry area at a temperature just above their minimum requirement. Tomatoes, for example, should be stored above 10°C (50°F). Store seeds from hardy plants in an airtight container in a refrigerator at 4°C (39°F) until spring.

Clean the flesh from fruit seeds by washing them in a sieve under the cold tap.

HOW TO SOW SEEDS

It's easy to germinate your collected seeds indoors or, if they are hardy, in a cold frame outside in early spring. Use biodegradable containers that can be planted together with the seedling's rootball in the soil, where they will decompose as the young plants grow.

YOU WILL NEED Harvested seed • Biodegradable pots or seed trays • Peat-free seed compost (or sieved garden soil) • Tray with transparent lid for plants that need heat • Watering can

1   Fill the pots or seed trays with compost or sieved garden soil. Press down lightly to remove any air gaps.

2   Sow the seed at the required depth. If you do not have the old seed packets to guide you, plant seeds at a depth of two or three times their diameter – for example, a pea seed should be planted 3cm (1¼in) below the compost. Tiny seeds usually need light to germinate and will require surface sowing, then covering with a very thin layer of compost.

3   Keep the seeds in a warm, light place or, if hardy, store them in a cold frame or sheltered spot in the garden. Place those that need heat to germinate in a propagator or in a tray with a lid on a warm windowsill. Water regularly.

4   Accustom the young seedlings to outdoor temperatures by placing the pots or trays outside during the day and bringing them in again at night for a couple of weeks before planting out. For tender plants, delay this until two weeks before the last frosts can be expected in late spring. This process is known as 'hardening off'.

TOP TIP IF THE SEED OF HARDY PLANTS FAILS TO GERMINATE, IT MAY NEED A PERIOD OF FROST TO BREAK DORMANCY. TRY PUTTING IT IN THE FREEZER FOR A COUPLE OF WEEKS BEFORE SOWING.

EASY SEED TO HARVEST

Alliums • French beans • Hollyhock (*Alcea rosea*) • Fennel • Sunflower (*Helianthus annuus*) • Poached egg plant (*Limnanthes douglasii*) • Honesty (*Lunaria annua*) • Love-in-a-mist (*Nigella*) • Peas • Peppers • Poppy (*Papaver*) • Pumpkins • Runner beans • Squashes • Tomatoes • Nasturtiums (*Tropaeolum*)

Sunflowers produce giant, flattish heads of large seeds that are easy to scrape off once the foliage is brown.

EASY PROPAGATION METHODS

It is simple to propagate new plants from those already in your garden. This allows you to multiply the resources available to pollinators and other wildlife in your garden at almost no cost to you or the environment. All the methods outlined here are quick and easy to do, although cuttings require more aftercare and a warm place in which to keep the propagated plants. Experiment to see what works for you, but always ensure that the plants you use are free of pests and diseases. If you want to take cuttings from a plant in a friend's garden, check first that it is not invasive or poisonous to wildlife.

Firm in your divided plants carefully to remove any air gaps.

DIVIDING PERENNIALS

Perennial plants can spread quickly through their root networks. After a few years, a plant may outgrow its space or the stems in the middle of a clump can become congested and will not flower as well as those on the outside. The solution is to divide the plant. In spring or autumn, water it well and trim back any long stems. Dig out the rootball and cut it into sections with a sharp knife or pull rooted segments apart with your hands. For large plants, insert two forks back-to-back in the middle of the clump and use them to prise the rootball apart. Discard old or unproductive stems in the middle of the clump and replant the divided sections in prepared soil. Water well during dry spells for the first year.

LAYERING SHRUBS AND CLIMBERS

This technique is ideal for climbing plants, such as clematis, Boston ivy (*Parthenocissus tricuspidata*), and climbing hydrangea (*Hydrangea anomala* subsp. *petiolaris*), as well as shrubs with flexible stems, including hazel (*Corylus*), daphnes, and flowering quince (*Chaenomeles*).

In spring or autumn, take a young, flexible stem, trim off side-shoots, and bend it to the ground. Where it touches the soil, remove the leaves from that section and use a sharp knife to make a slanting cut halfway through the stem, about 30cm (12in) from the tip. Insert a cane into the soil at this point and bury the wounded stem next to it. Tie the remaining stem to the cane. When it has rooted, you will see new shoots develop at the tip; at this point, you can cut the stem attached to the parent plant.

To propagate more than one plant, make a series of cuts along one stem, cutting just behind a leaf joint or bud. Then peg down the wounded areas, leaving the stems in between exposed to the light. New shoots will soon develop.

After cutting the stem, bury the wounded area under the soil surface. Insert a cane next to it and use soft twine to tie the stem to the cane.

TAKING CUTTINGS

Cuttings provide an easy way to create new plants from the current year's growth of many perennials and shrubs. Take them from vigorous side-shoots on healthy plants from early summer to mid-autumn. You can also make several cuttings from a long section of a year-old stem: trim each below a leaf joint at the base and just above one at the top. Cuttings are usually best kept indoors until they root.

1 Fill some containers with peat-free cuttings compost. Using clean secateurs, cut sections 7.5–10cm (3–4in) long from the tips of young, healthy, non-flowering shoots.
2 With a sharp knife, trim the stems just below a leaf joint and remove all the leaves except for a few at the tip.
3 Push the stems into a pot of compost, making sure that the leaves do not touch each other. Firm the compost around the stems and water gently.

Add a label and keep them at 15–21°C (59–70°F) in a propagator or tray with a clear lid. The cuttings should root within a few weeks.

AFTERCARE Keep the cuttings well watered but ensure the compost is not too wet, which may encourage fungal diseases. Pot them on when they have formed a few leaves and you can see roots through the holes in the bottom of their original container.

1

2

3

EASY CUTTINGS FROM SHRUBS

Roses and hardy deciduous trees and shrubs are easy to propagate from hardwood cuttings. In autumn, after the leaves have fallen, select straight stems that have grown during the current year. In a quiet area of the garden, make a narrow trench by pushing a spade into the soil. Cut the stems from the plant and remove any sideshoots and remaining leaves. Divide them into 25cm (10in) sections, making a straight cut just below a bud and a slanted cut just above a bud as you work towards the tip. Insert the cuttings into the trench, with the slanted end about 10cm (4in) above the soil surface. Firm in the soil around the cuttings and water well. Ensure the cuttings do not dry out. They will take about a year to root and produce new shoots.

Push the cuttings into a trench in the soil, slanted end upwards, and leave for a year to root.

TOP TIP WHEN TAKING HARDWOOD CUTTINGS OF TREES, LEAVE A SINGLE BUD ABOVE THE GROUND. FOR SHRUBS, LEAVE A FEW BUDS ABOVE THE SURFACE, EACH OF WHICH WILL DEVELOP INTO A SIDE SHOOT ONCE THE CUTTING HAS ROOTED.

BUYING NEW PLANTS

When you are looking for new plants from a nursery or garden centre, consider how and where they may have been grown before making a purchase. Locally grown plants will have a lower carbon footprint than those acquired from further afield or abroad. It is also worth asking what type of compost the plants have been grown in and making a few checks before buying to ensure your plants are healthy.

Asking garden centre staff about the company's policy on spraying plants with chemical pesticides may encourage eco-friendly methods.

BUYING LOCAL

We all need to buy new plants at some point, and the best way to minimize your carbon footprint in this respect is to purchase them from a local nursery. Look for suppliers near you who raise their own plants and can tell you where they bought their seeds or cuttings and whether they use peat-free compost and avoid chemical pesticides. Some nurseries also run pot-recycling schemes or supply plants without a plastic pot or bag, thereby reducing waste; ask about the packaging options when buying online too.

Large garden centre chains do not always have a wider choice of plants than small independent nurseries, and while they may not spray their plants with pesticides on their sites, their wholesaler or grower may have previously applied them. Plants that display the bee symbol, denoting their value to pollinating insects, may not have been exempt from spraying, either.

TOP TIP ALWAYS ASK RETAILERS WHICH COMPOSTS AND CHEMICALS HAVE BEEN USED FOR THEIR PLANTS. SUPPLIERS ARE MORE LIKELY TO RESPOND WITH ORGANIC OPTIONS TO MEET DEMAND.

NATIVE VS NON-NATIVE

While it might seem ideal to buy only native plants, they may not always be the most suitable. For example, the European native flag iris, *Iris pseudacorus*, would swamp a small pond, where the more compact, non-native blue flag *Iris versicolor* may be a better choice – frogs and toads will happily shelter among the foliage of either. Non-natives may also have been grown using organic methods in this country, which would mean they have a low carbon footprint, so ask the retailer where their plants were raised.

Iris versicolor is a compact marginal plant that is ideal for small ponds.

Iris pseudacorus is best for large wildlife ponds where it will be free to spread.

SWAPPING PLANTS

In addition to exchanging plants with friends, you may find that your local horticultural society has a plant-swapping scheme. There is no guarantee that the plants you receive will have been grown using organic methods, but they will usually have a low carbon footprint. Bear in mind that plants offered by other gardeners may spread quickly, which is why they have a surplus. Also take care if the original owner does not know the plant's species or name, as it may be an imported invasive type, or even a carrier of disease.

You may have surplus plants grown from seed that you can swap with friends. You can also make a plan with like-minded people in your area so that each of you grows something different, which can be exchanged within the group.

BUYER BEWARE

When buying plants or accepting them as gifts, make a few checks before taking them home. First, inspect the stems, flowers, and both sides of the leaves for signs of pests or diseases, and reject any plant you are unsure about to prevent importing problems into your garden. If you are able to, tip the plant gently out of its container and check that the roots are healthy, too. Also look for signs of poor growth – dry compost can indicate a lack of care, and multiple roots growing through the holes in the bottom of a pot is a sign that the plant is root-bound. Congested roots in a pot restrict growth.

Take time to examine a plant and the bottom of its pot before buying.

If possible, tip a plant out of its pot and check the condition of the roots.

WHAT IS BIOSECURITY?

With reference to plants, the term "biosecurity" describes the measures that are in place to protect them from harmful pests and diseases. These often arrive in a country on imported plants, which is another reason to buy from local suppliers who know the origins of their stocks. In Europe and the UK, all plants for planting and some seeds require passports when moved, which helps to prevent the spread of diseases such as *Xylella fastidiosa*. This devastating bacterial disease affects a wide range of species, including olive trees, lavender, and many herbaceous perennials. Similar restrictions are in place in the USA.

However small, an olive tree must be guaranteed free of disease.

NEED TO KNOW
- If there is no nursery selling organic plants nearby, try buying them online.
- Unpack plants as soon as they arrive, water them well, and keep them in a cool place before planting.
- Bare-root trees and shrubs, available in autumn and winter, can be planted temporarily in a trench or large pot in a sheltered spot if conditions are not immediately favourable for planting them in the ground.

PREPARING TO PLANT

The best way to ensure that your seeds germinate and plants thrive is to create the ideal growing conditions for them. Whatever your soil type, you can increase your plot's fertility using eco-friendly methods. You can also maximize the biodiversity and range of habitats in your garden by increasing your planting space with new beds and borders, and enlarging those you have already made.

Nutrient-rich, well-drained soil will help your plants to flourish.

MAKING A NEW BED

The best time to make a new bed is in autumn or early spring, when the ground is not waterlogged. Mark out an area with a hosepipe or pegs and string; squares, rectangles, ovals, and circles look neatest. To lift turf, use a sharp spade to cut it into small squares. Insert the spade under each one and slice through the grass roots before removing it. You can then compost the turf: set it upside down in a quiet area of the garden to decompose or use it to line the bottom of a raised bed.

Remove any large stones, building detritus or other rubbish from the soil. Also, dig out perennial weeds, such as bindweed, docks, and brambles, and hoe off annual weeds.

Alternatively, if the area is badly infested with weeds, put a layer of flattened cardboard packaging (remove any tape) on the surface and cover it with a 10cm (4in) layer of compost (see pp.64–65). Patience is required for this method as it will take about 12 months to kill off the weeds – the cardboard will decompose naturally.

Dig out perennial weeds and any other unwanted plants.

MINIMIZE DIGGING

Until recently, the advice to gardeners was to dig over the soil every year to incorporate organic matter, remove weeds, and reduce compaction. Now, research shows that minimizing soil disturbance is more beneficial to both the soil and the environment (see opposite). Huge volumes of carbon are locked up in the soil and digging it releases this greenhouse gas into the atmosphere, contributing to climate change. Turning the soil also breaks down its natural structure, which can increase compaction, reduce drainage, and lower its capacity to hold water and nutrients. Digging brings weed seeds up to the surface, too, where the increased light levels will trigger germination.

To lift squares of turf, slide your spade under each one, cutting through the roots as you do so..

THE NO-DIG METHOD

This method is not only more environmentally friendly, it also cuts down the amount of work you have to do, reducing weed growth and removing the need for digging. When you are making a new bed, simply add a 5–10cm (2–4in) layer of homemade compost or well-rotted manure over the surface (ensure you buy the latter from a reliable organic source – see p.51). On existing beds, apply the compost or manure each autumn, leaving a mulch-free area around the woody stems of trees and shrubs. Earthworms then pull it down into the lower depths of the soil, where they and microorganisms such as bacteria decompose it and recycle the nutrients it contains, ready for use by your plants. The thick compost layer also suppresses weeds by blocking the light they need to grow. Weed seeds will still germinate in the mulch layer, but its loose structure makes the seedlings easy to pull out. You can also plant your crops and flowers directly into the mulch; their roots will soon grow into the soil.

Applying a thick layer of manure or compost over the surface feeds the soil and helps to suppress weeds.

TOP TIP AVOID WALKING ON YOUR BEDS IF POSSIBLE SINCE THIS CAN CAUSE SOIL COMPACTION, ESPECIALLY IF THE GROUND IS WET. IN AREAS THAT REQUIRE FREQUENT ACCESS, SUCH AS VEGETABLE PLOTS, CONSIDER RAISED BEDS OR WALK ON PLANKS TO SPREAD YOUR WEIGHT. MULCHES WILL ALSO CUSHION THE LOAD.

A healthy population of worms is one of the most valuable components of productive soil.

WELCOME IN THE WORMS

The no-dig method encourages the activity of worms, which not only recycle nutrients from organic composts but also improve the soil structure. By burrowing into it they create air and water channels that improve drainage and sustain other organisms. Their mucus binds tiny clay soil particles together and clings to larger sand particles, creating a more stable soil structure; their casts are rich in plant nutrients and contain five times more nitrogen, seven times more phosphorus and 1000 times more beneficial bacteria than the original soil. Worm casts on the surface help to rebuild the topsoil layer, which is the most productive area.

CHOOSING COMPOSTS

Composts are invaluable to eco-gardeners. Some are ideal for growing plants in pots, others help to enrich and stabilize the soil in beds and borders. Compost produced from kitchen scraps and garden waste is the most environmentally friendly, but if you do not have space to make your own you can buy it from garden centres or online. However, commercially produced compost is less eco-friendly, especially if it contains peat, which is sourced from rare bog habitats. These important carbon sinks support plants and wildlife that are found only in their acidic conditions, so check products before buying.

Good-quality peat-free compost has an open texture that will allow good drainage.

COMPOSTS EXPLAINED

Whether homemade or bought, composts consist of organic material such as decomposed wood chips, leaves, plant stems, and vegetable matter. Homemade composts may also contain small quantities of cardboard and paper, while commercial types can include peat, coir, coconut fibres, sand, and grit. The most environmentally friendly composts are those you make yourself (see pp.52–53). Any that you buy will have a higher carbon footprint, and those containing peat should be avoided; others that contain materials that have not been sourced locally or are a byproduct of commercial agriculture may also be adversely affecting the environment.

PEAT-FREE COMPOSTS

Peat is sourced from peat bogs, which hold more than a quarter of all the soil carbon in the world, even though they cover just three per cent of the land area. Research shows that drained peat bogs and those cut to make compost and other products emit at least two billion tonnes of carbon dioxide into the atmosphere each year, while their destruction also damages the fragile ecosystems they support.

As an eco-gardener, you should buy only those composts labelled "peat-free" and made from products guaranteed free from pesticides and herbicides. Note that composts labelled "organic" do not necessarily meet these critera, so always check carefully before purchase. Soil-based composts, sometimes referred to as John Innes composts, will probably contain peat if not labelled otherwise. Bags of topsoil may also contain peat; ask suppliers for a list of the ingredients before buying.

You can make your own seed compost or buy commercial peat-free varieties.

Use homemade composts to raise ornamentals and crops in pots.

Horse manure from a nearby stable is an excellent source of fertilizer.

Having your manure delivered loose rather than in plastic bags from a garden centre is an environmentally friendly option.

WELL-ROTTED MANURE

Animal waste can be used as a soil conditioner, and will help to increase the fertility of your beds and borders in a similar way to compost. Dung is usually combined with straw and left to rot down, after which it can be used safely on your garden (fresh manure contains high levels of nitrogen, ammonium, and salts which can burn your plants). This traditional form of soil conditioner has been used for centuries, but today some animals are fed on grass or hay that has been treated with herbicides. The result is manure contaminated with harmful chemicals that may kill your plants, so check your sources to ensure their manure is from an organic farm or stable.

TOP TIP IF YOU ARE ABLE TO MAKE YOUR OWN COMPOST (*SEE PP.52–53*), YOU CAN USE IT TO CREATE POTTING COMPOSTS. COMBINE EQUAL AMOUNTS OF SIEVED HOMEMADE COMPOST, GARDEN SOIL, AND LEAFMOULD FOR A LIGHT AND NUTRIENT-RICH MIX SUITABLE FOR MOST CONTAINER PLANTS. FOR SEED COMPOST, TRY AN EQUAL MIX OF GARDEN SOIL, LEAFMOULD, AND HORTICULTURAL SAND.

MAKING LEAFMOULD

This crumbly material is a superb soil conditioner and can also be used as an ingredient for homemade potting compost. It is easy to make from autumn leaves that have fallen on paths, patios, or lawns – those that are on beds and borders can be left to rot down *in situ*. Pack the leaves into a sack or a simple cage made from wooden stakes and chicken wire or old pallets. Water the leaves if they are dry and place a stone on top of the sack or untreated wood on the cage to prevent them blowing away. Store for a year to make a leafmould mulch or up to two years for a finer material that can be used for potting composts.

Rake autumn leaves from lawns and hard surfaces and transfer them to a sack or cage to rot down into leafmould.

MAKING YOUR OWN COMPOST

Homemade compost is a wonderful, free resource for any eco gardener. It recycles kitchen and garden waste, transforming them into a nutrient-rich medium that promotes good soil structure and feeds hungry plants, delivering its goodness over a long period. You can buy or make your own compost bin, and it's easy to create this natural product, even in the smallest of gardens.

Beehive-style compost bins make decorative, practical features for a wildlife or ornamental flower garden.

Compost benefits all soils, improving their structure and releasing plant nutrients.

HOW COMPOST WORKS

As plants die, soil organisms, including bacteria, fungi, and worms, feed on the vegetation and break it down into compost. This organic matter is rich in plant nutrients and a spongy material called "humus". Humus helps to improve soil structure by binding together tiny clay particles to create channels between them that allow water to drain through more easily. It also absorbs large amounts of water, thereby making free-draining sandy soils more moisture-retentive.

CHOOSING A BIN

Compost heaps can be simple mounds of kitchen peelings and garden waste material covered with an old carpet or sheet of plastic, but the composting process will be much quicker if the material is better insulated. You can easily make a container yourself from old wooden pallets or recycled untreated wood, or even a large builders' bag, as long as it has drainage holes at the bottom.

Alternatively, buy a proprietary model. Wooden beehive-style bins look attractive in an ornamental or kitchen garden, while robust types made from recycled plastic, though less aesthetically pleasing, are inexpensive and will do the job just as well. Some bins also include a hatch that allows you to remove the compost from the base. If you have a large vegetable plot, consider two bins to ensure a plentiful supply.

Make sure your bin has a wide top to make filling easier, and a secure, rainproof lid that will not blow off. Set your bin on soil or grass — not on paving slabs — and position it where you can access it in all weathers and there is sufficient space for you to turn it and remove the contents.

Upcycled old pallets can be easily transformed into a practical compost bin.

Bins made from recycled plastic are inexpensive and easy to set up and use.

FILLING THE BIN

When filling your bin, aim for a mix of materials. Grass mowings, young weeds, soft plant stems, vegetable peelings, and flowerheads rot down very quickly and are known as "activators" or "greens". These are useful for starting the composting process, but will create a smelly wet mass if used on their own, so it is important to also add tougher plant-based materials, such as chopped or shredded twigs and prunings, old bedding plants, kitchen paper, and shredded cardboard boxes and toilet roll tubes. These woody items, known as "browns", decompose slowly but give the compost a good structure. Layer the materials, starting with a few twigs on the base to increase air flow and improve drainage, and then alternate soft greens with browns. Water the heap every 30–60cm (12–24in) as you fill it up. If your heap is too wet, add more browns; if it's too dry, add more greens or water. After 6–12 months, stop filling the bin and allow the composting process to finish.

Woody items, plants, and kitchen waste all contribute to a compost heap.

GREEN ITEMS TO COMPOST

Grass mowings • Annual weeds (but ensure they have not made seeds) • Raw fruit and vegetable kitchen waste, including coffee grounds and tea bags that do not contain plastic • Spent flowers and dead bedding plants

BROWN ITEMS TO COMPOST

Pet bedding, such as hamster, rabbit, or guinea pig hay and straw • Young hedge trimmings • Prunings – shred large branches first • Wood ash • Dry autumn leaves • Cardboard, egg boxes, paper towels, tissues, small amounts of newspaper • Shavings from untreated wood • Chicken, horse, or cattle manure if fed on organic grass

DO NOT COMPOST

Diseased plants • Perennial weeds • Meat • Fish • Cooked food • Coal or coke ash • Cat litter • Dog faeces • Paper items with plastic coatings

The lid keeps out rain and insulates the bin

Alternate layers of green and brown material

Add autumn leaves in thin layers or compost them separately

Put a layer of twiggy brown prunings at the bottom to aid air flow

Layering green and brown materials promotes a good structure and results in dark, sweet-smelling compost to use on the garden.

TURNING THE HEAP

To ensure your heap or bin works its magic, ensure there is plenty of air flowing through it. This sustains the micro-organisms that kick-start the composting process, as well as the larger creatures, such as beetles and worms, that work on it later as the contents cool down. Every few weeks, use a fork to turn over the layers to inject more air into the heap – you may notice that it heats up again after this as the micro-organisms start working again.

You will also notice that composting takes longer in winter when temperatures are low; continue to fill it at this time of year, but at a slower pace. Once the process is complete, you should be left with a dark, sweet-smelling, crumbly material that you can use as a mulch over your flower and vegetable beds, or to add to containers, mixed with some leafmould (see p.51) and garden soil.

Use a fork to mix the contents of your bin or heap to speed up the process and ensure all the material is fully composted.

GROWING YOUR OWN VEGETABLES

Food grown at home in a small vegetable patch or in containers on a patio or balcony will provide you with a supply of fresh, tasty produce with a very low carbon footprint. Sourcing your own vegetable seeds and using organic growing methods also means you can be sure that the food you are eating has not been contaminated by chemicals that could harm you or the environment.

Beetroot 'Boltardy' is a popular variety that is resistant to bolting.

A well-designed vegetable plot can help to feed your family all year round with nutrient-rich, organic produce.

WHERE TO GROW YOUR CROPS

Locate your plot on a reasonably flat patch of land in a sheltered spot. Most vegetables need plenty of sunshine, although some leafy crops will be happy in part shade. You won't need a huge space; in fact, a 1m (3ft) square plot will support a succession of tasty crops from early summer to late autumn, or you can grow compact crops in pots on a patio or balcony. It is a good idea to make your productive garden near a water source such as a water butt or outdoor tap, which will allow you to tend to your crops more easily.

Young crops need frequent watering, so locate them near a convenient source.

WHY GROW YOUR OWN?

Growing your own vegetables allows you to control the growing environment and maximize the benefits to you and to wildlife, particularly bees and pollinators that are affected by pesticides. Ensuring that your plot is free from harmful chemicals also protects the wider environment from contamination. In addition, growing food enables you to select varieties that contain more phytochemicals – which research shows have important health benefits – than the vegetables most commonly available in the shops (see p.56).

PREPARING YOUR PLOT

In the autumn, prepare an area for planting. First, remove large stones and dig out any pernicious weeds, such as brambles, dandelions, and bindweed, trying to remove the whole root system if possible. Hoe off any annual weeds too (see pp.64–65). Then cover the soil with a 5cm (2in) layer of home-made compost or well-rotted animal manure from a local supplier (see p.51). The compost or manure will be taken down into the soil by worms, which will release the nutrients it contains and help towards improving the soil structure.

Push a draw hoe through the soil to remove shallow-rooted annual weeds.

Vegetables grown in pots can be sited conveniently close to the kitchen.

PLANTING GREEN MANURES

If you have space for two plots, you can enrich the soil in one with nitrogen-fixing plants, known as green manures, while crops are growing in the other, then swap the crops and manures the following year. Green manures such as clover and beans can convert nitrogen, an essential plant nutrient (see p.63), from the air to maintain healthy growth, rather than taking it up in a solution from the soil like most plants. Sowing a patch with either of these plants in autumn and digging them into the soil in spring will release their nitrogen for summer crops to use. Flowering green manures, such as phacelia and crimson clover, can also be sown in spring after a winter crop has been harvested, and will provide insect pollinators with food before you dig them into the soil during the summer.

CROPS FOR POTS

Many crops grow well in large containers, such as half barrels or old dustbins. Fill them with peat-free compost and try a few of the crops listed below. Although large pots will need less frequent watering than smaller containers, it is still wise to set them close to a water supply to save you lugging heavy cans too far.

Try these crops for size:
Aubergines* • Beetroots • Carrots • Courgettes* • Herbs • Lettuce • Peppers* • Radishes • Spring onions • Swiss chard • Tomatoes*

*plant outside after the frosts

Crimson clover is an attractive and fast-growing green manure.

TOP TIP TO REUSE THE COMPOST IN A PRODUCTIVE POT, ADD SOME MORE COMPOST ON TOP OF THE EXISTING LAYER AND PLANT A DIFFERENT TYPE OF CROP THAT IS NOT SUSCEPTIBLE TO THE SAME PESTS AND DISEASES AS THE ORIGINAL, SUCH AS CARROTS FOLLOWED BY TOMATOES, OR COURGETTES FOLLOWED BY BEETROOTS.

PLANNING A VEGETABLE PATCH

Carefully planning your vegetable plot will pay dividends. For a succession of crops that delivers a steady harvest over many months, list the vegetables you want to cultivate and check their growing times. You can then combine a range of fast-growing crops during spring and summer that will produce more than one harvest with vegetables that take longer to mature. Experiment with crops you cannot buy easily in the shops, too, such as salsify or oca. You can make your own crop plan to include your favourite vegetable varieties or follow the one shown here, which is designed for beginners.

CHOOSING CROPS

You will want to grow the plants you enjoy eating, but to avoid disappointment, check that your garden conditions will suit them. Experiment with different cultivars for a variety of tastes and the maximum number of nutrients. For example, purple carrots are rich in anthocyanins, as well as the carotenes that are present in orange types, offering greater protection against cancer and inflammatory diseases than the vegetables that are widely available to buy. Other purple varieties, such as lettuces, onions, and cabbages, have greater health benefits than their green or white counterparts too.

Growing both purple and orange carrots offers you a variation in flavours.

Sow coriander seeds between rows of fast-maturing spinach and salad leaves, which will be harvested as the herb plants mature.

EXTENDING THE HARVEST

Fast-maturing crops, including salad leaves, salad onions, and radishes, can be squeezed between vegetables that take longer to mature, such as sweetcorn, cabbages, parsnips, and tomatoes. This practice, known as intercropping, is especially practical if you have only a small space for your vegetable garden.

Sowing small quantities of fast-maturing crops at six-week intervals also prevents gluts and extends the harvesting season. Remember that some vegetables are not hardy and should be sown indoors, then planted outside only after the last frosts have passed. These tender plants include tomatoes, courgettes, French and runner beans, cucumbers, and aubergines.

SIMPLE CROP PLAN

This simple crop plan is ideal for beginners as it includes vegetables that are easy to grow and provide a steady harvest from early summer to autumn. For more advice on growing each of the vegetables here, follow the instructions on the seed packets or check reliable sources online. If you do not have space for all the beds shown, simply pick out those with your favourite crops that will fit. You can also try intercropping the sweetcorn with beetroots and lettuces, as described opposite.

TOP ROW Sweetcorn × 9 • Summer squash × 1 • Runner beans × 8 • French beans × 6 • Maincrop potatoes × 4

MIDDLE ROW Dwarf bush tomatoes × 5 • Cucumbers × 3 • Kale × 4 • Courgettes × 2

BOTTOM ROW Beetroots × 20 • Carrots × 40 • Radishes × 40 • Swiss chard × 8 • Kholrabi × 12 • Oriental greens × 20 • Lettuces × 20 • Coriander × 6 • Parsley × 4

ROTATING CROPS

Try not to plant the same crops in the same place year on year, which can lead to a build-up of pests and diseases and depletes the soil of some essential nutrients. Vegetables fall into three main groups, and members of each have similar nutritional needs.

ROOTS AND LEAVES These include potatoes, beetroots, carrots, leeks, onions, Swiss chard, and spinach.

PEAS, BEANS, AND FRUITING VEGETABLES As well as peas and beans, this group includes tomatoes, peppers, chillies, courgettes, aubergines, cucumbers, and pumpkins.

CABBAGE FAMILY Known as brassicas, these include cabbages, kale, Brussels sprouts, turnips, radishes, broccoli, khol rabi, and oriental greens.

Ideally, use the same bed for roots in the first year, followed by peas and beans, which help to enrich the soil with nitrogen (see p.55) in the second year, and the cabbage family in the third year. Sweetcorn, herbs, and lettuces can be planted anywhere.

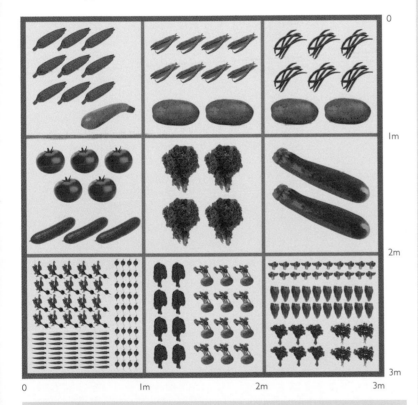

Plant Swiss chard with other leaves and roots such as onions.

NEED TO KNOW

- Keep beds weed-free (see pp.64–65).
- Start sowing tender vegetables indoors in spring; plant outside after the risk of frost has passed.
- Protect carrots, cabbages, and kale from insect pests (see pp.66–69).
- Make pyramid supports with canes for French and runner beans.
- Water crops during dry spells, focusing on vulnerable young plants.

GROWING YOUR OWN FRUIT

Growing fruit is one of the easiest ways to feed yourself and your family with delicious produce while helping to protect the environment. Trees and bushes fruit year after year with very little maintenance, and the flowers, foliage, and windfalls benefit pollinators and other wildlife in your garden. Fruits are also rich in phytochemicals, which help to protect us against illnesses such as cancer and heart disease.

Blueberry plants bear bell-shaped flowers before producing their fruit, classed as a "superfood".

Dwarfing rootstocks make it possible to grow several fruit trees in a small space.

WHERE TO BUY FRUIT TREES AND PLANTS

When buying trees and plants, look for growers who have organic certification. This ensures that their plants have not been sprayed with harmful chemicals and have been grown in peat-free compost, using organic methods. Most specialist nurseries give advice on the cultivation of the fruits they sell, or you can consult reliable sources online.

SELECTING FRUIT TREES

Tree fruits, such as apples, plums, and cherries, offer many benefits to wildlife, as well as helping to mitigate climate change (see pp.16–19). They are also suitable for small gardens, if you select those grown on dwarfing rootstocks; specialist nurseries can advise you on which will be most suitable for your plot. For example, apple trees are available on M27 rootstocks, which produce tiny trees suitable for growing in large pots on a patio; MM106 rootstocks give medium-sized trees suitable for a larger garden. Look out for heritage varieties to increase biodiversity, or do a little research to discover if any fruits were traditionally grown in your area, since these will probably thrive in your garden. Also check if your chosen tree needs a pollination partner tree to produce a good crop of fruit. Bare-root trees are cheaper than those grown in containers, but are only available to buy from late autumn to late winter.

If you want a single fruit tree of any type, choose a self-fertile cultivar such as pear 'Gorham'.

BOUNTIFUL BERRIES

In addition to a tree, plant a few fruit bushes, which take up less space in the garden but can be equally productive. Blackberries are very easy to grow, even in partly shaded areas, while raspberries and loganberries come a close second but will need full sun for their fruits to ripen. Plant blackberries and loganberries about 45cm (18in) from a wall or fence and train their stems along sturdy horizontal wires to create a fruiting screen. Raspberry canes are best planted in a bed or border and trained on a frame made from posts and wires, which will allow more airflow around the plants. If you have sufficient space, grow both summer- and autumn-fruiting varieties to extend your harvest.

You can also increase the biodiversity in your garden by selecting some of the more unusual fruit bushes, such as chokeberries (*Aronia*), which can be grown in part shade under a tree, along with decorative shrubs. Another berry to try is Chilean guava (*Ugni molinae*), which tastes a bit like a strawberry. This plant needs a warm, sheltered spot and acid soil, so grow it in a pot of ericaceous compost if yours is alkaline.

Chokeberries' black fruits against red autumn leaves offer ornamental value.

FRUITS FOR POTS

Where space is limited, grow fruits in containers. Strawberries will fruit well for two or three years in a hanging basket or pot. You can propagate them easily, too, by inserting the base of the little plantlets that form on long stems into individual pots of compost. Water them until they root, then sever the long stem to detach the new plant from its parent. Blueberries are also a good choice for pots, but you will need to grow at least two bushes to maximize your crop. These vitamin-rich berries need acidic soil, and containers allow you to grow them even if you do not have this in your garden. Plant them in organic ericaceous compost, such as products made from bracken and sheep's wool. Red and white currants will also grow well in a large pot and will produce long strings of nutrient-rich berries in summer. Try training the stems along wires fixed to a fence or wall so that the sun can reach all the ripening fruits.

Strawberries grow well in a basket, which can be hung at easy picking level.

Redcurrants are easy to grow and look attractive trained against a wall.

NEED TO KNOW
- Keep all newly planted bushes well-watered during dry spells for the first year. Water trees for up to five years until they are well-established.
- Make sure that your pots all have good drainage at the base; raise them off the ground on pot "feet" in order to allow water to pass through.
- You will need to water fruit growing in containers once or twice a week in dry or warm weather. Fit a rose head to a watering can and use at least one full can on each plant.

MAKING AN ECO-FRIENDLY LAWN

Much as we may love an expanse of green lawn, the environmental cost of maintaining closely mown, weed-free turf transforms what could be an ecological feature into a disaster zone. The effects of mowers, leaf-blowers, and lawn fertilizers create a toxic mix – one study has shown that petrol lawnmowers make up 5 per cent of the total air pollution in the USA. To reverse this trend and create a beautiful haven for garden wildlife, you can simply make a few adjustments to your maintenance routine, such as leaving the grass and flowers in your lawn to grow taller and bloom.

LEAVING YOUR LAWN TO GROW

While a carpet of closely mown turf makes a soft and natural ground covering in a garden, it offers very little value for the resident wildlife. Mowing it less frequently creates a window of opportunity for wildflowers such as clover, buttercups, daisies, greater plantain, and bird's-foot trefoil (*Lotus corniculatus*), which are present in most turf. If they are allowed to bloom, they will provide food for pollinators and the caterpillars of butterflies. Insects that breed in longer grasses also offer a feast for birds, as do the seeds of many lawn plants. In addition, clover enriches the soil with nitrogen, creating the ideal conditions for lush leafy growth.

If you must mow, leaving some of the clippings *in situ* will bolster growth – they release nitrogen as they decompose and can provide up to 30 per cent of the lawn's nutrient requirements. Add this to the contribution from clover, and the lawn will not need extra feeding. By mowing only once every three weeks you will greatly increase your lawn's wildlife benefits, while using a hand mower or scythe will lower your carbon footprint further. A lawn should also not need watering – turf is drought-tolerant, especially if the grass is longer, and it will soon revive when rain returns.

CONTROLLING WEEDS

Research has shown that the close cutting of lawns actually contributes to weed growth, because it exposes more soil between the blades of grass and so offers extra ground where seeds can germinate. Most flowers can simply be left to flourish in a longer lawn, but dandelions and dock can take over and decrease biodiversity. Dig some of them out with a knife or trowel or remove all of their top growth by hand rather than using chemicals, which harm wildlife and contribute to water pollution. If your lawn is too big to make hand-weeding a practical solution, consider converting it into a wildflower meadow.

Instead of fighting the weeds in a large lawn, allow some of them to bloom and contribute to the riches available to wildlife.

Remove pernicious weeds such as dandelions by hand before they spread.

TURNING A LAWN INTO A MEADOW

If you leave a patch of lawn to grow undisturbed for a season or two, neither feeding it nor using weedkillers, you will see flowers popping up among the grasses. Creating a small meadow in this way provides a habitat for insects and a refuge and food for birds and small mammals. It can take a few years to establish a good balance between the grasses and wild flowers, but over time a thriving ecosystem will form. Remove unwanted pernicious weeds by hand.

1 Some wildflower species will already be present in the grass mix. Grow more from seed in pots or buy them from specialist suppliers as small seedling plants (plugs). Annual yellow rattle (*Rhinanthus minor*) is particularly useful because it helps to reduce the vigour of established grasses, allowing more flowers to thrive. In spring, remove small sections of turf and plant or sow into the spaces.

2 For a more colourful meadow, sow annual wildflowers such as poppies, cornflowers, and corncockles into the soil between the grasses and perennial flowers in spring.

3 When the flowers have bloomed and set seed in late summer, lightly mow the meadow so that the grass is no less than 5cm (2in) high. Leave the cut stems *in situ* for a week to allow the flower seeds to disperse.

4 After a week, remove and compost the meadow clippings. If left in place they will rot down and nourish the soil, which will result in more grasses at the expense of wild flowers.

> **TOP TIP** IF YOU ARE LOOKING FOR A NATURAL GREEN CARPET THAT DOES NOT NEED MOWING, TRY A MICROCLOVER LAWN (*TRIFOLIUM REPENS*). IT IS DROUGHT-TOLERANT AND CAN BE WALKED ON, BUT REQUIRES RESOWING EVERY TWO OR THREE YEARS.

FEEDING PLANTS

There are various eco-friendly ways to feed your plants in order to keep them healthy and productive. Bulky organic fertilizers that are made from well-rotted animal manures and natural plant materials, such as homemade compost, release their nutrients slowly, allowing plant roots to absorb them when they are needed. They also improve the soil structure, so that it retains plant food and water while draining easily. Organic comfrey and nettle fertilizer teas are easy to make and offer another good source of nutrients for hungry plants, such as those grown in pots or fruiting crops.

Tomatoes are hungry plants and need plenty of potash to fruit well.

WHICH PLANTS NEED FEEDING?

In nature, plants that grow in one ecosystem generally share similar nutritional requirements. Conversely, a garden may have diverse species with very different needs; some may require an input of nutrients to grow successfully. Fruiting vegetable crops will require feeding, especially when grown in pots. Flowering plants, trees, and shrubs in beds and borders can also benefit from a 5–10cm (2–4in) layer of homemade compost or well-rotted manure spread over the soil surface each year in spring or autumn to keep plant nutrient levels topped up (see pp.48–53). However, a mulch may not be needed where the leaves that fall from trees and shrubs in autumn rot down and recycle their nutrients for the plants to take up.

FEEDING LESS FOR GREATER BENEFITS

Providing plants with plentiful nutrients may be counterproductive. For example, in species grown primarily for flowers, an excess of nutrients, especially nitrogen, may promote lush leaf growth at the expense of flower formation. Some crop species develop higher concentrations of phytochemicals (which help boost human immunity) when they are underfed. Excess nutrients are generally damaging to the environment, since they leach out of soils and accumulate in groundwater and natural waterways; so, unless your plants are showing signs of poor growth or have discoloured leaves, it is best to avoid artificial fertilizers.

Feeding crops less often and exposing them to full sun can increase their production of health-promoting compounds called phytochemicals.

PLANTS IN POTS

Plants that are grown in pots often suffer from more nutrient deficiencies than those grown in the ground. If they are in their pots for more than a year, remove the top layer of compost in spring and replace it with well-rotted homemade compost or manure. This will slowly release its nutrients as the seasons progress and should keep most plants healthy. Summer-flowering annuals and crops may need a boost from a homemade fertilizer, such as comfrey or nettle tea, to top up the nutrient supply. You can also buy seaweed extract, which is rich in potassium (see *right*) as well as trace elements such as iron that plants only need in small amounts. Check that any product containing seaweed is from a certified source.

Trees, shrubs, and other plants in their pots for more than a year require annual applications of compost or rotted manure.

KEY PLANT NUTRIENTS

There are three main nutrients that plants require. Acid-loving plants also need the trace nutrient iron.

NITROGEN (N) This nutrient promotes healthy foliage; a deficiency causes yellowing leaves and poor growth.

POTASSIUM (K) Also known as potash, potassium encourages flowering and fruiting and general plant hardiness. A deficiency causes poor flowering or fruiting and the leaves turn yellow or purple with brown edges.

PHOSPHORUS (P) This promotes good root growth. Deficiencies, which will cause poor growth, are rare, except in heavy clay soils or areas with high rainfall.

MAKING YOUR OWN FERTILIZER

Comfrey and nettles both make excellent natural fertilizers. Comfrey (*Symphytum officinale*) is rich in potassium (potash), while nettles have a high nitrogen content. Comfrey and nettle flowers are also a good food source for pollinators. To make these plant "teas", use 1kg (2lb) comfrey leaves to 15 litres (3 gal) of water, or 1kg (2lb) nettles to 10 litres (2 gal) of water.

1 Collect leaves and non-flowering stems, or harvest after flowering.
2 Chop up the leaves and stems and pack them into a bucket or other container with a lid. Weigh them down with a brick. Add water (preferably rainwater from a butt) and cover. Leave comfrey to ferment for about six weeks and nettles for three or four weeks.
3 Strain the liquid into a bottle with a cap – it is quite smelly – and store in a cool place. Dilute 1 part "tea" to 10 parts rainwater before applying to the soil or compost.

1

2

3

KEEPING WEEDS UNDER CONTROL

Weeds are vigorous plants that can take over a garden if left unchecked, which means you will have to remove some of them on a regular basis. Not all are pernicious, so be selective in how you tackle them. Some annuals, such as herb robert, provide food for pollinators and can be removed easily by hand if they threaten to overwhelm the garden, while nettles offer feeding grounds for some butterfly larvae.

Weeding by hand close to your plants avoids accidental damage from tools.

WHAT TO WEED

Weeds take light, nutrients, and water from other plants and can also reduce the biodiversity in your garden. While many weeds support wildlife, allowing invasive species to take over your plot will reduce the variety of food available and the range of creatures that visit it.

Plants to keep in check include the bramble, which spreads widely given the chance. Its sprawling roots throw up new plants, while birds eat the blackberries and disperse the seeds in their droppings. Brambles can also produce new plants from the tips of their arching stems, which take root when they touch the soil. If you want to feed the birds, opt for a blackberry cultivar such as 'Loch Ness' that will not grow as vigorously.

Dandelions are also best kept in check. Each plant produces up to 15,000 seeds, which soon form a blanket of flowers over the garden. While dandelions offer food to bees, there are less invasive plants that do the job just as well.

Other weeds can provide specific benefits to wildlife. Nettles, for instance, are the favourite food of red admiral, comma, and small tortoiseshell butterfly larvae, so just limit the spread of these beneficial weeds or set aside a "wild" corner (see p.122).

Dandelions seed themselves prolifically, so remove them as soon as they appear.

Dense planting reduces the open soil in which weeds can grow.

GROUND CONTROL

One of the best ways to keep weeds at bay is to cover your beds and borders with your chosen plants. Packing them in cheek by jowl creates a blanket of leaf cover over the soil, leaving little space or light for weed seeds to germinate. Just be vigilant in spring when perennial plants are emerging and the soil is exposed, and tackle any pernicious weeds as soon as you see them.

PERENNIAL WEED ATTACK

Bindweed, bramble, dandelion, couch grass, ground elder, and horsetail are among the most pernicious perennial weeds and the most difficult to eradicate. Never resort to chemical warfare, since weedkillers adversely affect pollinators, such as bees, and contaminate rivers and oceans; some studies also suggest that they may be carcinogenic. Such weedkillers are rarely a long-term cure anyway, because the weeds' roots may be spreading from a neighbouring garden that you cannot reach and seed will find its way in again regardless.

Instead, dig out these weeds, removing all the roots you can see. Place a marker where the weed was and keep checking the area for regrowth, digging it out as soon as it emerges. If you have a large patch of weeds, cover the whole area with an old carpet or other covering that will not allow any light through. Leave in place for at least 12 months, and the lack of light will kill the plants.

Where bindweed or dandelions are growing next to your plants, pull them out or cut off all the top growth – this may not kill the weeds completely, but it will weaken them and prevent them from stealing the light and nutrients your plants need.

Horsetail is a pernicious weed that will need to be dug out carefully to remove all the roots and prevent regrowth.

NEED TO KNOW
- Most perennial weeds grow from even a tiny piece of root, and some can produce roots from their stems too.
- Many small domestic compost heaps will not achieve the high temperatures needed to kill these weeds completely, so do not compost them.
- After pulling up perennial weeds, pack them in strong biodegradable bags and cover to exclude all light. Leave for two or three years and they should break down into usable compost.

HOEING SUCCESS

Remove any annual weeds, such as chickweed, groundsel, and fat hen, as soon as you spot them. Either pull out mature plants by hand before they flower or hoe off seedlings when they emerge. Get to know what each of these weeds looks like at different stages of growth so that you are able to identify them easily. On a fine day before the weeds come into flower, push a Dutch hoe blade just beneath the soil surface to sever the roots from the top growth. Leave the leaves and stems to wither *in situ* and then compost them.

Annual weeds can be easily removed by hoeing at seedling stage.

KEEPING PESTS AT BAY

Aphids, slugs, snails, and a number of other common garden pests can devastate crops and ornamental plants, but there are many ways to control them without resorting to chemicals that may also kill other forms of wildlife. One of the most effective methods is to encourage a range of pest predators such as frogs, toads, ladybirds, and hoverflies into the garden, where they will form an army of allies in your fight against pest attacks. Checking plants frequently for damage and limiting any infestations before they become a major problem are other eco-friendly ways to keep pests at bay.

Ferns do not attract many pests, but offer shelter for frogs and toads that eat them.

KEEPING PESTS AT BAY

Insects, birds, and small garden creatures can be friends or foes, depending on their food preferences. Rather than waging a constant war on pests, choose plants that will thrive in the soil and light conditions in your garden, since vigorous plants shrug off pest attacks much more effectively than struggling specimens. For minimum effort, include plants that few pests target. For example, hostas are beautiful plants for shade but slugs love them; instead, plant ferns, which like similar sites and are rarely troubled by slugs, nor any other pests. In sunnier sites, snails and slugs adore dahlias, lettuces, and basil, so try sedums (*Hylotelephium spectabile*), Swiss chard, and thyme, which are more resistant to attack.

ENCOURAGING PEST PREDATORS

One of the best forms of defence against plant pests is to lure their predators to your garden. Birds are omnivores and some, including thrushes and blackbirds, eat snails, while sparrows feed aphids to their young in spring and early summer. Encourage them into the garden with supplementary food, such as fruit trees and shrubs with berries, as well as seed feeders.

Frogs and toads are your friends, too. They eat slugs, snails, and insects, helping to keep many pests at bay. Both require some water, and leafy plants and piles of stones to hide beneath. Even a small garden can accommodate them – a half barrel filled with water and pots of marginal plants will provide a home for frogs, as long as they can hop in and out easily.

The larvae of hoverflies, lacewings, and ladybirds enjoy a diet of aphids and will quickly polish off infestations, as will adult ladybirds. These beneficial insects will soon find their way into gardens where aphids are on the menu. Hoverflies and lacewings are also pollinating insects, so attract them with pollen-rich plants.

Adult ladybirds will seek aphids and other small insect pests to eat.

Ladybird larvae are predators of aphids and may eat a hundred per day.

PROTECTING THE YOUNG AND VULNERABLE

Seedlings and young plants are particularly vulnerable to attack from pests such as slugs and snails, though many that suffer damage at this early stage will not be affected so badly as they mature. To protect young plants, pot on seedlings and keep them indoors or out of harm's way until they have developed sturdy stems. This may mean growing vulnerable plants such as cosmos in pots until they are almost in flower. Remember to put them outside during the day for a couple of weeks before planting them out permanently after the frosts. Also mix up your planting with different species and varieties – many pests have plant preferences, so the damage will be limited if they cannot easily hop on to their favourites nearby.

Protect young plants indoors until their stems are more resistant to attack.

Bright red lily beetles are easy to spot and pick off vulnerable plants.

KEEPING WATCH

You will soon become familiar with the pests that are attracted to the plants in your garden. Check vulnerable plants every day or two and remove any pests that have few predators, such as the red lily beetle. Be vigilant for vine weevils, too. These slow-moving beetle-like insects are easy to catch as adults – they make semi-circular holes in leaves and do little damage, but their grubs eat plant roots, often with fatal results. Slugs and snails may be too numerous for pest predators to control completely. During the day, look under leaves or stones where they may be hiding; you can collect them and place them near your pond to feed amphibians and birds.

POND PESTS

A pond with plenty of frogs, toads, and newts should not suffer with too many pests, as these predatory amphibians will keep most under control. Water lily beetles and china mark moth caterpillars carve holes out of the foliage of pond plants but they rarely spoil the flowers, so it is usually best to just tolerate this minor damage. Most water snails help to keep the water clear, but the great pond snail, which has a long, pointed shell, will eat pond plants. If these arrive in your pond, attract them with lettuce leaves and remove them with a net.

The great water snail may arrive in your pond by means of eggs on aquatic plants.

PEST DEFENCE

Where neither predators nor being vigilant succeeds in keeping pests at bay, you can use biological controls and other eco-friendly methods to protect your plants. Food crops, such as cabbages, carrots, and soft fruits, are particularly vulnerable to insect and bird damage and will often need special treatment. Some tolerance of birds eating your tree fruits is recommended, especially as there is usually enough for everyone, but it can be disheartening to find they have polished off your entire crop of juicy raspberries before you felt they were ready to pick.

Snails and slugs can rapidly cause major damage to many types of plant.

BIOLOGICAL CONTROLS

The term "biological controls" refers to tiny pest predators that you usually buy in packets and apply to vulnerable plants. They do not disrupt natural ecosystems and are safe to use on organic crops. The most popular are nematodes – microscopic creatures that eat a range of pests – and parasitic wasps that consume aphids. Neither will harm other wildlife in your garden, but check that you buy the correct type for the pest you want to control. There are specific types of nematodes that prey on slugs or vine weevils, for example, or mixes that will eat fruit flies, carrot root fly, onion fly, gooseberry sawfly, and codling moth. Store packs of biological controls in the refrigerator until you are ready to use them and read the application instructions carefully. Most work best from spring to early autumn when the pests are active. You may have to apply biological controls a few times during the growing season (for example, slug nematodes last up to six weeks) and they can be quite expensive.

Biological controls, such as those that kill onion fly, are applied by mixing with water and dousing vulnerable plants.

Nasturtiums planted with broad beans will draw aphids away from your crop.

SACRIFICIAL AND COMPANION PLANTS

Where crops are besieged with a particular type of pest, you can sow another plant that the pest prefers to act as a decoy. A good example is to plant a frill of nasturtiums (*Tropaeolum majus*) around broad bean plants to attract aphids to their flowers rather than your crops. Other plants can be used to lure predators that will feed on pests. Pot marigolds (*Calendula*) act as both a sacrificial and companion plant, leading aphids away from crops and attracting hoverflies and lacewings that prey on them. Other good companion plants that attract these beneficial insects are French marigolds (*Tagetes*), mint (*Mentha*), and sunflowers (*Helianthus annuus*), as well as flowering crops such as celery, dill, and parsnips. There is little scientific evidence that some plants, such as mint and sage, deter pests by confusing them with their scents, although many gardeners claim they provide good protection.

BARRIER METHODS

Many pests can be prevented from ruining your crops by means of simple barriers. A fruit cage or netting that keeps birds away is a good investment if you want to grow soft fruit, such as berries; when buying netting, check that the packaging states that it is harmless to birds. Likewise, fine netting over a crop of carrots will prevent carrot root flies from laying their eggs in the soil, while covering cabbages and other brassicas will offer protection against cabbage white butterflies.

The cabbage root fly can also be kept off your crops by a protective collar fitted closely around the stems of the plants. Cut out circles 10–15cm (4–6in) in diameter from cardboard or recycled plastic and make a straight cut to the centre so that you can tuck the card or plastic around the plant stem and prevent the fly from laying her eggs next to the crop.

PEST TRAPS

Beer traps have been used for centuries to control slugs and snails and offer cost-effective protection. Bury a jam jar or similar container in the soil with the rim level with the surface and fill it with beer. Molluscs are attracted to the brew and drown when they fall in. You can also try placing inverted pots stuffed with straw on canes pushed into the ground. These will lure earwigs away from vulnerable plants such as dahlias.

Beer traps for slugs are easy to set up by sinking a container into the soil.

Wool-based barrier products can help to deter slugs and snails from attacking young crops. Grit or coffee grounds are also worth a try, as these pests are said to dislike the texture and will not traverse them, although the protection they afford is limited and some will venture over them to the treats beyond.

A netted cage will prevent a crop of soft fruit from being eaten by birds.

NEED TO KNOW
- Some organic insecticides made from natural materials such as pyrethrum (derived from chrysanthemum and *Tanacetum coccineum* plants), oils, or soaps can help to control insect pests, such as aphids and thrips.
- Organic pesticides tend to be non-specific, which means they will also harm beneficial insects, such as hoverfly and ladybird larvae. Use them cautiously.

DEALING WITH DISEASES

One of the best forms of defence against disease is to provide your plants with the conditions they need to thrive. Many diseases can be kept at bay by ensuring your soil is fertile and has good structure and plants are not overly stressed by drought or nutrient deficiencies. You can also select disease-resistant crops that offer some immunity and grow a wide variety of plants to prevent infections from spreading too quickly. When problems do arise, identify the disease and take action – but never resort to chemical controls, many of which have a detrimental effect on wildlife and the environment.

HOW DISEASES SPREAD

Many diseases are spread by plant pests, such as aphids and other sap-sucking insects, so take action to reduce vector populations (see pp.66–69). Some weeds also carry disease – for example, groundsel acts as a host for rust spores that may then infect your crops or ornamental plants. Keep tools clean and always wipe the cutting blades after use on each plant with hot water and soap, which will kill most diseases and viruses. Crop rotation (see pp.56–57) will also help to prevent the spread of diseases that afflict specific plants, such as potato blight and onion rust.

Groundsel often carries rust disease, so weed it out to prevent transmission.

Magnesium deficiency causes inter-veinal yellowing of leaves.

DISORDERS VS DISEASE

Sometimes a sickly-looking plant may be suffering from a disorder rather than disease, and these problems are usually easy to solve. Wilting plants, for instance, may simply need a good watering, while plants with uncharacteristic yellow or red leaf tints may require feeding. Try these remedies first before considering diseases.

COMMON PLANT DISEASES

When watering and feeding do not cure a problem, check the plant for signs of disease and then act promptly. The following common ailments affect a wide range of plants and crops.

DOWNY MILDEW *Symptoms*: leaves are discoloured on the upper surface and show white, grey, or purple mould below. *Control*: remove infected parts. Spores are transmitted during wet weather and when humidity is high. Water in the morning so leaves dry out during the day, avoid overhead watering, and improve ventilation around plants.

GREY MOULD *Symptoms*: this fungal disease (*Botrytis cinerea*) causes fuzzy grey-brown mould on decaying leaves, stems, flowers, and fruits. Buds and flowers may shrivel and die. *Control*: it is often caused by high humidity, so provide better ventilation and drainage. Remove dead plant material promptly when it drops on to the soil or compost.

HONEY FUNGUS *Symptoms*: this disease kills the roots of perennials, trees, and woody plants. White fungal growth that smells of mushrooms appears between the bark and wood, usually at ground level, and honey-coloured toadstools may also appear. *Control*: there is no cure, but to limit its spread, create a vertical physical barrier from pond liner, setting it at least 45cm (18in) deep in the soil, with 2.5cm (1in) above the surface. This will prevent the underground thread-like fungal strands (rhizomorphs) from spreading.

POTATO AND TOMATO BLIGHT *Symptoms*: tomato and potato leaves have a white fungal growth on the undersides, and shrivel and turn brown. Brown patches appear on tomatoes, while potato tubers turn reddish-brown beneath the skins and succumb to rot. *Control*: do not compost infected material – bin or burn it. Practise crop rotation and buy disease-resistant varieties. Indoor tomatoes are rarely affected.

POWDERY MILDEW *Symptoms*: white powdery fungal growth covers the leaves, flowers, and fruits, which may then become distorted. *Control*: remove infected parts and improve the air flow between plants by spacing them more widely. Add mulches over the soil to retain water, but also ensure that drainage is good to prevent other fungal diseases taking hold.

RUST *Symptoms*: raised areas, mainly on lower leaf surfaces, which may be rusty brown, orange, yellow, black, or white, depending on the specific type of rust. Heavy infection reduces the vigour of some plants; in others, it has little effect on fruiting and flowering and will not need any management. *Control*: where infection is localized, remove infected material but do not take off too many leaves, because this will affect growth. Remove dead and diseased plant material lying on the soil in autumn and do not compost.

VIRUSES *Symptoms*: pale green or yellow spots, streaks, or mosaic patterns appear on the leaves. Flowers are smaller than usual and may be streaked with white patches; fruit may also be discoloured and streaked. *Control*: remove and destroy infected material; do not compost or the virus will spread. Remove weeds and control pests, such as aphids, which carry viruses. Wash your hands after handling infected plants.

Grey mould appears on all parts of plants and may cause them to die.

Blight affects the leaves, fruit, and tubers of tomatoes and potatoes.

Powdery mildew is identifiable by a white powdery growth on plants.

PEST- AND DISEASE- RESISTANT PLANTS

Whether you are new to gardening or just want to add extra colour and interest, try a few of these plants, which rarely succumb to pest or disease attacks and need no special treatments. Most thrive in a wide range of sites and soils and provide beautiful flowers and foliage, as well as benefits to wildlife.

ALLIUM 'PURPLE SENSATION'
ALLIUM HOLLANDICUM 'PURPLE SENSATION'

HEIGHT AND SPREAD 90 × 20cm (36 × 8in)
SOIL Well-drained
HARDINESS Fully hardy
SUN ☼ ☼

In late spring, this ornamental garlic produces tall stems topped with spheres of tiny purple flowers. The leaves appear first but fade as the flowers open, so plant it between small shrubs and perennials that will hide the dying foliage. Pollinators love allium flowers so this bulb is a must for any wildlife garden, although it also looks good in more formal settings. Plant in groups of five or more in autumn in free-draining soil for the best visual effects.

The round flowerheads of this allium are magnets for bees and other pollinators.

CHINESE ANEMONE *ANEMONE HUPEHENSIS*

HEIGHT AND SPREAD 90 × 60cm (36 × 24in)
SOIL Well-drained/moist but well-drained
HARDINESS Fully hardy
SUN ☼ ☼

An easy-to-grow perennial, Chinese anemone produces lobed mid-green leaves and abundant open flowers held on tall, sturdy stems from midsummer to early autumn. The flowers of the species are pale pink, but there are purple, white, and dark pink cultivars, most of which are not quite as vigorous. This anemone soon forms large clumps, but it can be easily controlled by digging out unwanted plants. Use it in a wildlife garden or in informal mixed beds to attract a range of pollinators.

'Hadspen Abundance' is a cultivar that bears a profusion of deep pink flowers.

SNAPDRAGON *ANTIRRHINUM MAJUS*

HEIGHT AND SPREAD 90 × 40cm (36 × 16in)
SOIL Well-drained
HARDINESS Hardy down to -5°C (23°F)
SUN ☼

With hundreds of cultivars to choose from, this annual offers something for everyone. The unusual flowers, which open like dragon's jaws when pinched, come in colours from white and yellow to pink and red. They are magnets for pollinators and are produced continuously throughout the summer and early autumn, if plants are dead-headed regularly. Sow seed in pots indoors in spring and plant outside after the frosts in beds or pots in a sunny spot; snapdragons will also flower in a little light shade.

Brightly coloured flowers appear in clusters from summer through to autumn.

MICHAELMAS DAISY *ASTER × FRIKARTII*

HEIGHT AND SPREAD 70 × 40cm (28 × 16in)
SOIL Well-drained
HARDINESS Fully hardy
SUN ☼ ☼

This easy-going perennial produces slim green leaves and daisy-like lavender-blue flowers on tall stems from midsummer to mid-autumn. These are rich in pollen and nectar, providing food for bees and many other pollinating insects. One of the most popular cultivars is 'Mönch', which is disease-resistant and very long-flowering. Plant it in groups of three or more for a dramatic effect. If the stems flop, stake them with bamboo canes when the new growth appears in spring.

'Monch' has an extended flowering season which makes it ideal for a wildlife garden.

PURPLE BERGENIA *BERGENIA PURPURASCENS*

HEIGHT AND SPREAD 40 × 40cm (16 × 16in)
SOIL Moist but well-drained
HARDINESS Fully hardy
SUN ☼ ☼

The large, glossy, evergreen leaves offer cover for hibernating insects and garden creatures in winter, when the foliage also takes on dark red tints. The leaves provide a foil for the late spring flowers, which rise above them on sturdy stems and attract pollinators. Most bergenias are easy to grow, given average soil and a few hours of sun each day. Plant in groups at the edge of tree canopies or at the front of a mixed flower border, where the leaves will add a decorative frill throughout the year.

The handsome leaves of this bergenia create a subtle backdrop for the pink flowers.

ENGLISH MARIGOLD *CALENDULA OFFICINALIS*

HEIGHT AND SPREAD: 30 × 20cm (12 × 8in)
SOIL Well-drained
HARDINESS Hardy to -15°C (5°F)
SUN ☼ ☼

The sunny orange or yellow flowers of the English marigold are set off by slender mid-green leaves. Guaranteed to cheer up any patio container or garden border, the blooms of this hardy annual also attract pollinators. For more subtle cream-coloured flowers, choose a cultivar such as 'Snow Princess' or 'Lemon Twist'. Sow seed in spring *in situ* or in pots indoors or outside, and use it en masse at the front of a border, in a herb garden – the petals are edible – or in containers.

The cheerful flowers of English marigold brighten any border or container.

TRAILING BELLFLOWER *CAMPANULA POSCHARSKYANA*

HEIGHT AND SPREAD 15 × 45cm (6 × 18in)
SOIL Moist but well-drained/well-drained
HARDINESS Hardy to -15°C (5°F)
SUN ☼ ☼

This bellflower produces a carpet of small, heart-shaped green leaves on long, wiry stems and small blue starry flowers which bloom from summer to early autumn. However, its spreading habit means that it can overwhelm less vigorous plants, so allow it plenty of space to ramble. A favourite with bees and other pollinators, it can be planted in cracks in paving, to tumble over a wall, in a large rock garden, or at the edge of a tree canopy. You can pull out some stems in spring to keep its growth in check.

The bellflower's starry blue flowers are a reliable source of colour over a long season.

PERENNIAL CORNFLOWER *CENTAUREA MONTANA*

HEIGHT AND SPREAD 45 × 45cm (18 × 18in)
SOIL Well-drained/moist but well-drained
HARDINESS Fully hardy
SUN ☼ ☼

Also known as mountain knapweed, this robust plant thrives in all but the harshest conditions. It produces lush silvery-green leaves and spidery blue flowers with purple centres from late spring to midsummer. The white cultivar 'Alba' is a little less vigorous than the species. This cornflower is beautiful in a wildlife garden or informal border where bees and other pollinators will be drawn to the flowers. It is inclined to spread, so give it space to form a clump and pull out unwanted seedlings in spring.

Famed for their colour, the flowers of perennial cornflower appear over a long period.

MOUNTAIN CLEMATIS *CLEMATIS MONTANA*

HEIGHT AND SPREAD up to 6 × 5m (20ft × 16ft)
SOIL Well-drained
HARDINESS Fully hardy
SUN ☼ ☼

In spring, the spiralling stems of this vigorous climber are covered with mid-green leaves and fat buds that open to reveal four-petalled white flowers. Cultivars in shades of pink are also available. The plant offers a wonderful encore when grown through a fruit tree, the pollen-rich flowers taking over the show as the tree blossom starts to fade. It can also be trained on horizontal wires fixed to a boundary fence or wall and it needs no pruning, apart from a trim after flowering, if needed, to keep it in check.

A profusion of white flowers decorates this clematis in spring alongside pretty lobed leaves.

MALE FERN *DRYOPTERIS FILIX-MAS*

HEIGHT AND SPREAD 90 × 90cm (36 × 36in)
SOIL Moist but well-drained/moist
HARDINESS Fully hardy
SUN ☼ ☼

Most ferns demand damp soil to thrive, but the male fern is happy in drier conditions, making it useful for shady areas beneath trees. In the spring its fronds unfurl to form a shuttlecock of finely divided mid-green leaves. They may overwinter or turn bronze as temperatures fall, but they will look tatty by the spring and will need removing to make way for new growth. Grow this fern in large clumps for a textured effect, or with spring bulbs such as bluebells that will bloom before the leaves have fully expanded.

The finely divided foliage of the male fern offers shelter for garden birds and other wildlife.

HARDY FUCHSIA *FUCHSIA MAGELLANICA*

HEIGHT AND SPREAD 1.5 × 1m (5 × 3ft)
SOIL Well-drained
HARDINESS Hardy to -10°C (14°F)
SUN ☼ ☼

Also known as lady's eardrops, this deciduous shrub may be evergreen in mild areas and city gardens. The woody stems bear many small green leaves and the dainty scarlet and purple pendant flowers, which resemble earrings, appear from summer until mid-autumn. The nectar-rich flowers are loved by bees, butterflies, and moths. Cultivars come in a wide range of colours, from white to pale pink and dark red. Plant it in a sheltered spot in a mixed border or, in milder areas, use it as a hedge.

The pretty red and purple flowers of this fuchsia dangle like earrings from the leafy stems.

SWEET WOODRUFF *GALIUM ODORATUM*

HEIGHT AND SPREAD 20 × 35cm (8 × 14in)
SOIL Well-drained/moist but well-drained
HARDINESS Fully hardy
SUN ☀

The dainty foliage and flowers of this woodland plant belie its tough nature. It quickly makes a large clump, spreading via rhizomes (underground stems) to produce a carpet of bright green divided foliage. From late spring to summer, starry, scented white flowers appear, attracting bees and other pollinators. A good choice for groundcover beneath trees, woodruff can be combined with spring bulbs such as narcissi and scillas, but keep it in check by removing stems in spring if it starts to swamp its neighbours.

Small, starry white flowers stand out against the foliage, attracting pollinators,

BLOODY CRANESBILL *GERANIUM SANGUINEUM*

HEIGHT AND SPREAD 30 × 40cm (12 × 16in)
SOIL Well-drained/moist but well-drained
HARDINESS Fully hardy
SUN ☀ ☀

Producing a low hummock of small, lobed leaves, bloody cranesbill explodes with colour from late spring to midsummer, its masses of magenta cup-shaped flowers almost obscuring the foliage. Use it at the front of a border to suppress weeds or in raised beds, where it can tumble over the sides. Cut back the old flowerheads and leaves in summer to encourage a fresh flush of growth and divide large clumps in spring. The foliage provides cover for insects and small garden creatures.

This cranesbill's magenta flowers provide bright colour over several weeks.

HELLEBORE *HELLEBORUS*

HEIGHT AND SPREAD up to 60 × 60cm (24 × 24in)
SOIL Well-drained/moist but well-drained
HARDINESS Fully hardy
SUN ☀

Hellebores are essential for winter and early spring gardens, blooming when many plants are still under the soil. *H. niger*, the white-flowered Christmas rose, appears from mid- to late winter, followed in early spring by *H.* × *hybridus*, *H.* × *ericsmithii*, and their many cultivars in colours ranging from dark purple and pink to cream and green-tinged white. All offer pollinators food as they emerge from hibernation. When the flowers appear, you can cut off faded leaves to keep the plant looking neat and make way for new foliage.

The large white flowers of *H.* × *ericsmithii* are tinged with pink and borne on pink stems.

ICE PLANT *HYLOTELEPHIUM SPECTABILE*

HEIGHT AND SPREAD 45 × 45cm (18 × 18in)
SOIL Well-drained
HARDINESS Fully hardy
SUN ☀

The fleshy grey-green leaves of this perennial emerge in spring, followed in late summer by dusky pink flowerheads which bloom into autumn, providing late nectar for bees and butterflies to sustain them through hibernation. The bronze seedheads remain throughout winter, adding a beautiful feature to the garden. The ice plant works well at the front of a sunny mixed border, in prairie-style schemes, and in wildlife gardens. To prevent stems flopping, cut one stem in three back to the ground in early summer.

The flowerheads of the ice plant offer late-season sustenance for pollinating insects.

OREGON GRAPE 'CHARITY' *MAHONIA × MEDIA*

HEIGHT AND SPREAD 3 × 3m (10 × 10ft)
SOIL Well-drained/moist but well-drained
HARDINESS Hardy to -15°C (5°F)
SUN ☼ ☀

The Oregon grape is an architectural shrub with long stems of spiny evergreen leaves that offer year-round interest. From late autumn to winter, it bears spikes of cup-shaped small yellow flowers which have a sweet scent and attract pollinators. These are followed in spring by blue berries. Useful for shady borders and areas under trees, this plant makes a textural backdrop to spring bulbs, ferns, and smaller shrubs such as daphnes. Keep it in check by cutting back long stems in spring after flowering.

The bold and striking leaves of the Oregon grape are notable all year round.

DWARF CATMINT *NEPETA RACEMOSA*

HEIGHT AND SPREAD 60 × 60cm (2 × 2ft)
SOIL Moist but well-drained
HARDINESS Fully hardy
SUN ☼ ☀

This free-flowering perennial produces clumps of slender stems covered with aromatic foliage, topped in early summer with spikes of small, lilac-blue flowers. The foliage is loved by cats, while the blooms are a magnet for pollinators, including bees and butterflies. 'Walker's Low' is a popular cultivar, with compact growth and abundant blooms. It is an excellent plant for the front of a border in an informal or wildlife garden. Cut back the stems after the first blooms have faded to promote a second flush in early autumn.

Catmint's myriad lilac-blue flowers are covered with bees when they appear in summer.

LOVE-IN-A-MIST *NIGELLA DAMASCENA*

HEIGHT AND SPREAD 45 × 40cm (18 × 16in)
SOIL Well-drained
HARDINESS Fully hardy
SUN ☼

The finely dissected foliage and blue flowers surrounded by feathery bracts make this eye-catching hardy annual a must for any cottage or wildlife garden. The flowers open in summer, when they attract bees and other pollinators, and the seedheads that follow are equally decorative, persisting for many weeks into early autumn. Sow the seed *in situ* in spring, after which, given free-draining soil and a sunny site, this little annual will self-seed in future years. The blooms also make beautiful cut flowers.

The pretty blue and white flowers of love-in-a mist are followed by attractive seedheads.

COMMON LUNGWORT *PULMONARIA OFFICINALIS*

HEIGHT AND SPREAD 30 × 30cm (12 × 12in)
SOIL Well-drained
HARDINESS Fully hardy
SUN ☼ ☀

This semi-evergreen perennial has white-spotted hairy leaves that create a textural carpet throughout most of the year in sheltered areas. In spring, clusters of pink tubular flowers appear, then change to blue, giving a two-tone effect. They are loved by bumblebees and other pollinators. Perfect for a woodland scheme or the front of a shady border, this plant needs little maintenance, but it is best to remove old leaves in spring to make way for fresh new growth. Also cut back the flowering stems after the plants have bloomed.

With pink and blue flowers simultaneously, lungwort brings valuable colour to a shady spot.

ROSE 'SEAGULL' *ROSA 'SEAGULL'*

HEIGHT AND SPREAD up to 6 x 5m (20 x 15ft)
SOIL Well-drained/moist but well-drained
HARDINESS Fully hardy
SUN ☼

A vigorous rambling rose, 'Seagull', like many of its kind, is generally free of pests and diseases. Tall climbing stems hook on to supports to reach the light. In spring, mid-green glossy leaves appear, followed in summer by large clusters of small, fragrant, pure white single or semi-double flowers with yellow stamens. The blooms, which attract pollinators, are followed by small red hips. Use 'Seagull' to decorate a tree or pergola, or tie it to sturdy wire horizontal supports on a house wall.

Bearing pretty white flowers with yellow stamens, 'Seagull' is ideal for growing through a tree.

ROSEMARY *SALVIA ROSMARINUS*
(SYN. *ROSMARINUS OFFICINALIS*)

HEIGHT AND SPREAD 1 x 1m (3 x 3ft)
SOIL Well-drained
HARDINESS Hardy to -10°C (14°F)
SUN ☼

Rosemary is a shrubby evergreen herb, loved for its aromatic needle-like foliage, used in culinary dishes, and its small blue pollen-rich flowers, which appear in spring. Grow it in the herb garden, or in mixed, sunny borders or large pots. Trailing varieties, such as *S. officinalis* (Prostratus Group) 'Capri', look good spilling over the sides of raised beds. Pruning is not necessary, but cutting stems for cooking from spring to late summer will encourage bushier growth.

Edible aromatic foliage makes this plant ideal for both kitchen and garden.

PINCUSHION FLOWER *SCABIOSA CAUCASICA*

HEIGHT AND SPREAD 60 x 40cm (24 x 16in)
SOIL Well-drained
HARDINESS Hardy to -10°C (14°F)
SUN ☼

The perfect plant for a cottage-style or wildlife border, this elegant scabious produces slim, grey-green leaves and wiry stems of lavender-blue flowers with pincushion-like centres, which appear over a long period in summer. Rich in pollen, they attract a range of butterflies, bees, and other pollinators. They also make good cut flowers. Plant it in groups at the front of a border or mix with meadow grasses for a naturalistic effect. Removing faded blooms will extend the flowering period.

The lavender-blue flowers of this scabious are popular with many insect pollinators.

ZINNIA *ZINNIA ELEGANS*

HEIGHT AND SPREAD 90 x 30cm (36 x 12in)
SOIL Well-drained
HARDINESS Not frost-hardy
SUN ☼

Zinnias' tall stems of brightly coloured flowers are favourites with florists and also make a striking feature in the garden. This annual produces mid-green leaves along its stems and flowers in a range of colours, including pink, white, orange, and yellow. To attract pollinators, select the species or cultivars with open flowers – some are also short and more compact, ideal for the front of a flower bed. Sow seed in spring indoors and plant the young plants outside after the frosts in large pots or sunny borders.

The orange flowers of *Zinnia marylandica* 'Double Zahara Fire' bloom from early summer.

Making biodegradable seed pots from recycled newspaper is easy and free – they can be planted together with the seedlings and will quickly decompose in the soil.

REDUCE, REUSE, AND RECYCLE

Buying new garden products and furniture is sometimes necessary, but you may find recycled or salvaged items do the job just as well. By reusing and recycling, you limit the non-biodegradable waste that is burned or goes to landfill, while also reducing the plastic pollution that blights our world. The ideas in this chapter show you how to give household products another life in the garden and ensure your patios, boundaries, and furniture are eco-friendly.

REDUCING AND REUSING PLASTICS

Plastic is so ubiquitous in the horticulture trade that it is difficult to buy a plant which will help to protect the planet without also acquiring a plastic pot. While some nurseries and garden centres run recycling schemes, it is still all too easy to end up with a mountain of plastic waste that will take more than 400 years to break down. Plastic is also responsible for a cocktail of polluting chemicals that are pumped into waterways and the atmosphere during its manufacture and as it degrades. So, here are some easy ways to reduce, reuse, or even eliminate the plastic in our gardens.

PLASTIC TYPES

Many of the plastics that are used in the horticulture trade are not easily recycled. Until recently, black low-grade polypropylene (PP) was widely used for pots. This cannot generally be recycled because the sensors at recycling facilities do not recognize the black pigment. Although plastics of a different colour are now more prevalent, they still have a high carbon footprint, generating pollution during their manufacture.

Plastics are identified by a number, and these are the most common types found in gardening products

4 LDPE (LOW DENSITY POLYETHYLENE) Used in potting compost bags and polytunnel covering, this plastic is rarely recycled at commercial facilities.

5 PP (POLYPROPYLENE) Ropes, netting, and rigid containers such as flowerpots and plant trays are often made from this plastic. It can be reused and recycled, unless it is black.

6 PS (POLYSTYRENE) The International Agency for Research on Cancer has identified styrene as a possible human carcinogen. Used in plant packaging, it is not generally recyclable.

Many everyday household containers can be repurposed for use in the ecological garden.

REUSING PLASTICS

There are billions of plastic pots and other products used for gardening and food packaging that are destined to sit for hundreds of years in landfill or be burned in incinerators that churn out toxic air pollutants. Some will end up in our rivers and oceans, too, so we need to reduce our dependence on plastic products and repurpose the plastic we already have. Before buying plastic seed trays, look through your refrigerator and store cupboard for alternatives. Yoghurt pots make excellent seed and plant pots (see p.83) and are often made from plastic that can be used several times, unlike some flimsy plastic module trays that split after the first sowing. Trays used for soft fruits and vegetables such as mushrooms can also be repurposed, while small black plastic plant pots that cannot be recycled are ideal for sowing seeds and potting on seedlings. After use, wash them in soapy water and they will be ready for the following year.

SWAP SHOPS

Look online for nurseries near you that run plant-pot recycling schemes, where you return their pots and they wash and reuse them. If your local nursery or garden centre does not offer such a scheme, ask them if they could introduce one – consumer pressure often results in positive action. Also enquire about pot-swap schemes, where you bring your own pots or bags to take home your chosen plants.

Take your own container to pot swaps to avoid bringing more plastic home.

BYPASSING PLASTIC

To prevent bringing more plastic into your garden, select bareroot trees and shrubs rather than container-grown plants, and ask the seller if they can package them for transport to your home in a biodegradable material, such as newspaper or cardboard.

You can also buy biodegradable seed and plant pots made from plant fibres, husks, or animal dung, known as cowpat pots. Some are reusable; others you plant in the ground where the pots will decompose naturally – breaking them open carefully before planting helps the roots to grow out more easily. Alternatively, make your own from paper or other materials (see pp.82–83).

Traditional terracotta pots, which will last a lifetime if you look after them carefully, offer another eco-friendly option. They are made from natural clay, and if you buy them from a local potter, thereby reducing packaging and transportation, they will have a very low carbon footprint. Also look out for clay pots in antique shops and ask friends if they have some that they don't need lurking at the back of a shed.

A biodegradable, compostable plastic-free cup made from plant material.

WHAT ARE BIOPLASTICS?

Made from renewable resources such as vegetable fats and oils or plant starch, bioplastics use fewer fossil fuels in their manufacture and degrade much more quickly than ordinary plastic. Some are also compostable. Plant pots made from these materials are now available to buy.

Plant-fibre pots are better for the planet and blend well in the garden.

Terracotta pots can be used many times and have a low carbon footprint.

MAKING YOUR OWN SEED POTS

There is no better way to raise your seeds than in pots and seed trays that you have made yourself from recycled and biodegradable materials. Some can be planted along with your seedlings, where they will decompose and help to feed the soil as they do so. They are also fun to make with children and take only a few minutes – less time, in fact, than a visit to the garden centre to buy commercially produced pots.

Food cans with drainage holes in the base make eco-friendly seed pots.

Cardboard rolls from paper towels are the perfect size for germinating seeds.

ROLL OVER

Possibly the easiest seed pots you will ever make, the cardboard rolls inside toilet paper and kitchen towels are perfect for deep-rooted seedlings, such as sweet peas (*Lathyrus odoratus*) or runner and French beans. You can use rolls from toilet paper whole and cut those from kitchen towels in half. Pack them in a rigid container, as they tend to wobble a bit and may lose some of their structure when wet. Fill with compost and start sowing. The cardboard is biodegradable and can be planted in the ground or a large container.

HATCHING SEEDLINGS

Cardboard egg cartons make excellent ready-to-use, biodegradable seed trays, and the eggs themselves can also provide a home for seeds. If you are partial to a boiled egg for breakfast, do not discard the shells. Make a small hole in the base with a skewer, fill them with compost and sow one or two seeds into each shell. You can plant the shells along with their seedlings, but crack them carefully first so that the young roots are able to escape more easily. As the shells decompose, they release calcium, an essential plant macronutrient that is needed for building cell walls. It also helps to maintain a healthy chemical balance in the soil.

Sowing seeds in eggshells avoids plastic and feeds developing seedlings.

YOGHURT PLANT POTS

Plastic food containers of many types can be upcycled into seed trays and pots. Yoghurt pots are particularly useful – the small types are ideal for individual seeds, while the larger containers are perfect for potting on seedlings. Clean the pots thoroughly in hot soapy water first to remove all traces of food. Heat up a kitchen skewer and use it to make two or three drainage holes in the base of each yoghurt pot, then fill with compost and sow your seeds or transplant your seedlings.

Small yoghurt pots are convenient for sowing individual seeds.

MAKING POTS FROM NEWSPAPER

These eco-friendly seed pots are fun to make with children. You can use newspaper, comics, or magazines, but avoid embossed paper, which may contain plastics or harmful chemicals. Use your pots for large vegetable seeds such as beans or cucumbers and annual flower seeds, including cosmos, annual dahlias, zinnias, and calendulas. Do not leave the pots in a tray of water, as this may cause the paper to decompose too quickly.

YOU WILL NEED Newspaper, comic, or magazine paper • Small can or glass jar • Scissors

1 Flatten a sheet of paper then fold it in half lengthways so that you have a rectangle comprising two layers of paper.
2 Lay a small glass jar or drinks can on its side and place it on one end of the folded paper. Roll the paper around the jar or can a couple of times.
3 Fold the newspaper at one end over the bottom of the jar or can to form the base of the pot. Carefully remove the newspaper pot and stand it up on a table.
4 Trim the paper at the top so that the pot is the height you want it, plus about 2cm (¾in). Fold this excess paper inside the pot to secure it. You are then ready to fill your pot with seed compost (see p.51) and sow your seeds.

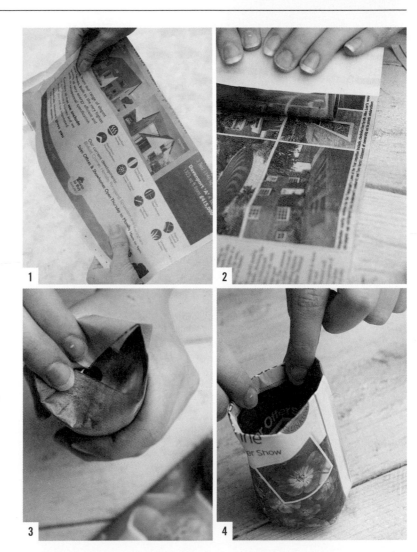

TOP TIP SOW JUST ONE OR TWO SEEDS INTO EACH HOMEMADE PAPER POT SO THAT YOU WILL NOT NEED TO TRANSPLANT THE SEEDLINGS INTO LARGER CONTAINERS BEFORE TRANSFERRING BOTH POT AND PLANT INTO THE SOIL OR POTTING COMPOST.

ECO-FRIENDLY LANDSCAPE MATERIALS

Most gardens need hard surfaces, fences, and walls. Concrete appears to provide an inexpensive solution here, but its manufacture is the cause of great environmental damage. Decking and fences made from natural wood are a good alternative, if sourced from sustainably managed woodland local to you, but in many cases, using products made from recycled materials, even plastics, may be your best option.

Steps made with compacted soil, gravel, and recycled timber lend a natural look that is well suited to informal planting.

REUSING PAVING

Paving made from concrete, brick, and natural stone has been used for centuries to create paths, patios, and terraces. However, concrete manufacture releases huge amounts of CO_2 into the atmosphere, and making clay bricks requires burning fossil fuels, creating other noxious gases as a by-product. While natural stone is of course a naturally occurring material, its extraction and transportation can have a high environmental cost.

To mitigate the polluting effects of all these products, look for recycled options rather than buying them new. Repurpose old concrete slabs, which you may be able to source for free on local recycling websites – even broken pavers can be made into decorative crazy paving. Recycled bricks and natural stone products are widely available from salvage companies. They may be cheaper than new stock, and they also come with an aged patina that looks more natural in a garden setting than a brand-new material.

Repurposed materials are often easy to source and give the impression of a long-established garden.

Sustainably sourced timber is an eco-friendly form of decking.

DECKING OPTIONS

A wooden deck can be an eco-friendly option when choosing hard ground cover for your garden, but check that the timber has not been cut from a rainforest or other unsustainable source. Unless it is from a local woodland, it may have a significant carbon footprint resulting from its transportation.

A deck made from recycled plastic may also be ideal for an eco-garden, as more than 90 per cent of plastics are not repurposed and end up in landfill, or polluting the countryside, rivers, and oceans. Some composite decking uses wood reclaimed from manufacturing industries as well as plastic. These decks last longer than soft wood and require no chemical treatments, but check that the products you buy are made from 100 per cent recycled materials.

NEED TO KNOW
When buying wood, check for accreditation that proves it is from a sustainable, regulated source. In the UK, look for wood products that have certification from the Forest Stewardship Council (FSC) or the Programme for the Endorsement of Forest Certification (PEFC). These bodies assure that all wood and wood-based products originate from sustainable sources.

Recycled scaffolding boards and timber have been used to make these sustainable steps and deck.

Logs used for screens or barriers provide homes for wildlife.

REUSING MATERIALS FROM THE GARDEN

The most eco-friendly materials may be free and right outside your back door. Logs from pruned trees make beautiful stepping stones through gravel or grass, or use them to create screens that double as homes for hibernating insects and other wildlife. Prunings are ideal for homemade picket-style fencing, or you can harvest stems from a clump of bamboo to create an eco-friendly screen – if you cut just some of the stems, the plant will soon recover.

RECLAIMED TIMBER

Most timber waste produced by the construction industry goes to landfill, where it decomposes, releasing sequestered carbon back to the atmosphere as methane, one of the most damaging greenhouse gases. However, a wide variety of recycled wood for decking boards, paths, raised beds, and screens is available from reclamation companies and by buying it you increase the perceived demand, which may result in more wood being recycled in the future.

When purchasing reclaimed timber, check that it has not been treated with creosote or chromated copper arsenate (CCA), both of which pose risks to human health, wildlife, and the environment. These chemicals are now banned in most countries.

Crushed shells from the seafood industry make a biodegradable garden path.

ECO-FRIENDLY AGGREGATES

Aggregates made from the waste products generated by various industries, such as shellfish processing and glass manufacturing, provide a beautiful surfacing solution from materials that would otherwise be discarded and go to landfill. Look online for a range of products, from recycled television and computer screens to slate and terracotta chips – the byproducts of tile and china clay manufacturing.

CHOOSING AND MAKING FURNITURE

New garden furniture is often made from mixed materials that are difficult to recycle. A more sustainable solution is to give new life to used furniture and save it from landfill, perhaps by adding a lick of non-toxic paint or rubbing down the wood to refresh it. You can also lower your carbon footprint by making your own furniture from eco-friendly materials, or asking a local craftsperson to make it for you.

With some DIY skills and a few tools you can make your own classic table and bench set from recycled wood.

UPCYCLING SECOND-HAND FURNITURE

A great way to reduce your carbon footprint is to recycle used furniture. Search freecycle websites, auction houses, and charity shops for wooden tables and chairs – many will be a fraction of the cost of new items. Even plastic furniture will be better for the planet if it is reused rather than going to landfill. You may need to replace the cushions for chairs and benches, or simply re-cover the originals with organic cotton and wool or recycled materials such as curtains. Don't dismiss timber furniture designed for the home – if it is made from solid wood, it should last a few years outdoors, or longer if it is made from hardwood. To give old pieces a quick make-over, try painting them with a non-toxic, water-based paint or apply natural wax.

Cast iron furniture is long-lasting and eco-friendly. Vintage pieces are available online, at auctions, and in antique shops, but finding a complete dining set can be difficult, so mix styles for an eclectic look. A coat of non-toxic, solvent-free metal paint can help to unify mismatched items.

Secondhand metal furniture sourced from auction houses and antique shops will add a stylish period look to your garden.

Interior furniture that is no longer needed indoors, such as this classic wooden chair, can be given a new life outside.

MAKE YOUR OWN

If you would like to take a creative approach to your garden furniture, try making your own from upcycled or recycled materials. One of the simplest options is to use cut sections of logs as stools. Ask a tree surgeon, the local parks authority, or an estate manager in your area if they have any suitable logs to spare; never take them from woodland. You may need to get help from a tree surgeon to cut the logs to the correct lengths. Another easy project is to transform a cable reel into a stool or table with a little eco-friendly paint; these reels are available from auction and freecycle websites or directly from the manufacturers.

To create a bench for an informal garden, try screwing together a few wooden pallets, available from DIY shops, garden centres, and online. Rub down the surfaces with sandpaper to remove any splinters and sharp edges, and use wood offcuts to screw two pallets together, one on top of the other, to create the base. Screw a third pallet to the seat to make a back for

the bench, and paint it using a non-toxic product. Finally, add some cushions for extra comfort.

It is also relatively easy to make a classic timber garden table and bench set, if you have the space and tools for the job. To make the tabletops and bench seats, use metal joining plates or wooden battens to screw together recycled timbers cut to length. Then attach the legs securely to each piece with long coach screws.

Cushions can be among the least eco-friendly products in your garden if the covers are made from synthetic or new cotton fabrics and they are filled with polyurethane foam, a plastic polymer derived from oil. Look for products made from sustainable bamboo, organic or recycled cotton and wool, or recycled bottle plastic. Use these to cover existing cushion pads or buy those made from natural sustainable kapok, wool, buckwheat seed, or millet husks.

| **TOP TIP** MANY SOFT-FURNISHING PRODUCTS ARE NOT WATERPROOF. MAKE SURE YOU HAVE SPACE TO STORE THEM INSIDE WHEN THEY ARE NOT IN USE.

Old pallets can be screwed together and painted to create a stylish bench. Add some cushions for colour and comfort.

Logs from sustainably managed woodlands make eco-friendly seats and a wildlife habitat too.

Furniture made from offcuts from the timber trade has a low carbon footprint.

HANDCRAFTED ITEMS

Furniture-makers can offer beautiful items hand-crafted from natural materials, such as locally sourced or reclaimed wood and canes. Ask them where they buy their materials to make sure they are from certified sustainable sources (see p.85). Craftspeople may offer a selection of ready-made items or you can commission bespoke pieces to suit your needs. Either way, locally produced furniture that is made to a high standard using low-energy methods will have a much lower carbon footprint than products manufactured on an industrial scale.

Also look out for specialist companies that produce garden furniture from recycled plastics and other eco-friendly materials. You will find chairs made from 100 per cent recycled plastic recovered from the ocean, hammocks created with organic, recycled cotton fabric, cushions sewn from recycled plastic bottles, and beanbags that are hand-crafted from old yacht sails that would normally end up in landfill. Shipping distances will add to the carbon footprint of these products, but the total may not be as high as that of other bought goods.

RECYCLED PLANTERS

Look around your home, shed, and garden for items that are no longer needed for their original purpose and can be transformed into planters. Almost any vessel that has not been treated with toxic chemicals can be adapted, and finding a new use for bags, pots, and even old furniture will save them from landfill or the incinerator. When filled with plants, they can create intriguing focal points on a patio or in the garden.

CHOOSING CONTAINERS

Many household items can become eco-friendly planters: plastic food tubs, metal saucepans, washing-up bowls, and shopping bags will all do the trick. Look for items of furniture, too, such as old wooden drawers and cupboards. A tin bath will home long-lasting perennials or small shrubs, while a hessian bag that will biodegrade is best suited to short-term annual crops — you can add it to the compost heap when it is no longer useful. Your containers must have drainage to save plants from drowning, so make a few holes at the base of each item, using a drill with a metal bit for tough plastic and metal containers.

Stacks of bricks provide an informal support for recycled containers.

Sedums, houseleeks, and pretty pebbles create an unusual chair seat.

MAKING A PLANT SEAT

An unusual way to display succulents and other shallow-rooted plants is to create a plant seat in an old wooden chair. You can replace the original seat base with a piece of pond liner or an old plastic compost bag. For a neat finish, position the black side of the bag facing downwards, allowing enough depth of plastic to provide room for compost and plants. Fold the plastic bag or pond liner over to create a double layer before nailing it in place around the edge of the chair frame, then make some holes in the base for drainage. Add gritty compost for low-growing succulents such as sedums and houseleeks (*Sempervivum*), or sow baby salad leaves into a mix of potting and seed compost. If you would prefer wild flowers, make the base a little deeper.

An old apothecary's chest makes a characterful container for plants.

An old handbag can become a home for crops such as chard and spring onions.

HOW TO MAKE A COLANDER BASKET

Old colanders make decorative hanging baskets for alpine strawberries, salad leaves, or pollen-rich annuals. If you do not have one tucked away at the back of a cupboard, they are easy to find in charity and junk shops. As they have drainage holes already, all you need are some old chains or strong twine to hang them up once planted.

YOU WILL NEED An old colander • Recycled plastic (optional) • Chains and wire or strong twine • Potting compost (see pp.50–51) • Plants • Hook or metal eye

1 Punch two or three drainage holes into some recycled plastic and line the base of the colander to help it retain more water. (If you would prefer not to do this, you will need to water your colander more frequently.) Then half-fill it with potting compost.
2 Add your plants and fill in carefully around the root balls with more compost. Press it down lightly with your fingers to remove any large air gaps.
3 Thread the wire through three evenly spaced holes at the top of the colander and use it to attach the chains. Bring the chains together at the top and make a loop with more wire to connect them. Alternatively, attach the chains to a hook or metal eye. If using twine, thread it through the holes and tie it at the top.
4 Water the plants well and hang the colander from a hook, metal eye, or hanging basket bracket. Make sure you can reach it easily to water it. Water the plants every other day, or every day during hot weather.

DISPLAYING YOUR PLANTS

You can raise the stakes by creating a plant pot display on an old wooden ladder – just place your pots on the steps so that each plant receives sufficient light for its needs. Setting plants on old wooden chairs or recycled timber planks on bricks are other quick and easy ways to create a plant display on a patio or balcony.

Presenting plants on the steps of a ladder allows them all plenty of light.

RAISED BEDS FROM RECLAIMED TIMBER

Vegetables and soft fruits may grow more successfully in a raised bed than on open ground, especially if your garden has poor or heavy clay soil. A bed crafted from reclaimed timber and filled with soil and garden compost will provide your crops with the optimum growing conditions – and while homemade beds may be no cheaper than kits bought from a garden centre, the environmental cost may be much less.

Planting in raised beds allows you to reach your crops from adjacent paths, without treading on the soil.

GROWING IN RAISED BEDS

There are many crops and flowers that will thrive in a raised bed. Try growing salad leaves, onions, radishes, cabbages, and peas, plus tender crops such as runner beans, tomatoes, courgettes, and sweet peppers. Including a few pollen-rich flowers, such as French and pot marigolds, will brighten up the appearance of the beds and attract bees to help set the fruiting crops.

The richer soil conditions in a raised bed allow you to grow crops closer together than you would in the ground. This not only makes the most of your productive space but also helps to reduce the number of weeds, because the dense leafy growth cuts out light and prevents their seeds from germinating. Crops can be protected from birds and flying insect pests by inserting canes into each corner of the raised bed and covering it with appropriate netting (see *pp.68–69*).

Higher raised beds reduce the need for stooping to tend and harvest the crops.

CHOOSING MATERIALS

Salvage companies offer reclaimed scaffold boards and timbers ideal for making a raised bed. Railway sleepers are sometimes used, but those that have been treated with creosote, which contains toxic chemicals and is banned in most countries, are not recommended for eco-gardens or growing crops. Products will vary considerably in price, depending on the treatment they have undergone; for example, old floorboards may need nails to be hand-pulled from them, but scaffolding boards generally require less work to prepare them for sale. Check the credentials of your supplier, and ask about the wood's provenance, especially if it is hardwood that may have come from a rainforest.

Filling your raised beds with rich soil allows you to grow your crops closer together.

MAKING A RAISED BED

This practical bed should not take more than a day to make if you have some DIY skills. You can opt for a bed with just one layer of timbers, but adding a second layer to create a taller bed gives a more comfortable height to tend.

YOU WILL NEED Heavy-duty gloves • 8 timbers cut to size required: those used here are 4 x 2m (6ft) and 4 x 1.2m (4ft) • Sticks or chalk • Spade • Builder's spirit level • Tape measure • Rubber mallet • Drill • Screwdriver • Long heavy-duty coach screws • Mixture of topsoil and garden compost

1 Lay out the timbers on the ground to form a rectangle and mark their positions with sticks or chalk. Move the timbers aside, then use a sharp spade to remove turf and weeds from the areas on which they will sit, as well as any pernicious weeds from the bed area. You do not need to remove grass from the bed as the soil or compost will cover and kill it.

2 Place the first layer of timbers in position. Check that they are level by laying a builder's spirit level diagonally across the top; remove or add soil beneath them as required. Also make sure the base is square by checking that the diagonals are equal lengths.

3 Use a rubber mallet to adjust the position of the timbers so that they butt up and align neatly at the corners. Drill two holes on one side of each corner and secure the joints using coach screws. Check that all timbers are screwed together firmly.

4 Arrange the next layer of timbers so that the joints at the four corners are staggered, as shown, to give extra stability. Check that all the levels are correct before screwing the timbers together; there is no need to screw the second layer to the first as the timber is heavy enough to remain in place. Fill the bed with a mixture of topsoil and compost.

SECURING TALL BEDS

You may need to make a bed more secure if it has three layers of timbers or you are using narrow scaffold boards. Screw an upright piece of wood to each timber inside the frame at all four corners. For extra drainage and to reduce the amount of soil needed to fill the bed, add a layer of broken pots or builder's rubble to the base.

Adding upright struts at the corners gives a raised bed extra stability.

CHOOSING TOOLS AND EQUIPMENT

Garden tools made from recycled or biodegradable materials such as wood will often outlast cheap plastic items, and many can be repaired rather than sent to landfill when parts break. Secondhand cutting and digging tools, even those with plastic-coated handles, can last for decades, so look out for old items that are still serviceable. Taking good care of your tools and storing them in a dry shed will extend their life further.

Storing tools in a clean, dry shed or garage will increase their longevity.

Metal rakes and hoes with wooden shafts can last a lifetime.

Sharpening pruning saw blades regularly keeps them in working order.

REPAIRING TOOLS

If you buy tools with wooden shafts, they can be repaired rather than discarded should they break. You will find replacement wooden handles in various sizes and styles for spades, forks, rakes, and hoes online and at DIY and hardware stores. Online videos show how to replace a broken handle, but if you do not have the equipment or skills for this, ask a local professional to do the job. Once removed, an old handle can be cut into sections to make a log pile or hotel for hibernating insects and small garden creatures (see pp.124–125).

Replacing broken wooden handles on old tools extends their lives.

BUYING TOOLS TO LAST

Basic tools such as spades, forks, trowels, and hoes are essential items for all gardeners. As with any product for an eco-garden, check the quality and the materials used to make them before buying. Tools crafted from sustainable wood (see p.85), metal, bamboo, and cork can last a lifetime, given a little care, so before buying new, look for heritage items that are still in good condition. It may not be possible to restore old cutting tools with badly rusted blades, but those with a thin layer of rust can easily be brought back to life (see opposite). If you do buy new tools, check the length of the guarantee, which usually reflects their durability and quality.

WATERING EQUIPMENT

When buying a new hose, check that it is labelled "lead-free" – many older hoses contain lead in their brass fittings and in the hose itself, and should be avoided. Vinyl hoses made from PVC (polyvinyl chloride) contain phthalates, which have been linked to problems with human reproductive development and some cancers, so to be on the safe side choose a natural rubber hose with stainless steel fittings. Also check with the manufacturer that the rubber is from a sustainable source. If you cannot find one that fits the bill, it may be best to irrigate your plants in another way, while reducing the water needed by covering the soil with mulches. Many plastic watering cans are made from PVC, so opt for galvanized metal types, which tend to have a longer lifespan too.

Before buying a hose, check that it is free of both lead and phthalates, and the rubber is from a sustainable source.

RESTORING GARDEN TOOLS

To remove rust from old cutting tool blades, soak them in a 50:50 solution of vinegar and water for 24 hours, then remove any residual rust with wire wool. Wipe the blades dry and apply some linseed oil to protect them from rusting again in the future.

To maintain secateurs and other cutting tools, clean and sharpen the blades regularly – blunt blades can tear your plants and introduce disease. Wipe the blades with linseed oil and rub them with sandpaper to remove dried sap and stains. With a small sharpening stone, also known as a whetstone, gently rub along the blade. Match the angle of the stone to the tool's blade to create a sharp edge. If the blades of your secateurs break, you can buy replacements.

Keep the wooden handles of your tools in good condition by cleaning them with water and a stiff brush. If they become scuffed or splintered, use sandpaper to smooth them and then apply a little natural linseed or coconut oil with a soft cloth.

TOP TIP YOU CAN GET TOP-QUALITY, INEXPENSIVE SECONDHAND TOOLS ON FREECYCLE AND ONLINE AUCTION WEBSITES AND SAVE THEM FROM GOING TO LANDFILL. MOST ARE MADE FROM MIXED MATERIALS AND CANNOT BE RECYCLED EASILY.

Remove rust from cutting blades by rubbing them with wire wool.

Sharpen your secateurs by rubbing a whetstone along the blades.

After cleaning your tools, apply oil to the wooden parts with a soft cloth.

MAKING ECO-FRIENDLY EQUIPMENT

Before buying any equipment, look for items at home that can be repurposed for use in the garden. Products such as cold frames and plant supports can be expensive and new equipment may include non-biodegradable components, so making your own garden essentials is a great option to save money and reduce your carbon footprint. Packaging, pieces of wood, and kitchen utensils can be used to protect and cultivate plants, while larger items, such as panes of glass to cover tender plants or barrels for collecting rainwater, can be sourced from freecycle websites and neighbourhood swap schemes.

EASY MAKES

It is easy to make a few practical gardening aids, even if you have little space and no specialist DIY tools or skills. For example, you can cut the bottoms off plastic bottles and use the top halves as mini cloches to protect vulnerable seedlings in spring. To scare birds away from newly sown seedbeds or crops, such as cabbages, try stringing up strips of used aluminium foil or place old plastic bottles on bamboo canes, so that they rattle in the breeze. Alternatively, make a traditional scarecrow with prunings for arms, old clothes for the body, and a sock stuffed with straw or newspaper for the head.

A homemade cold frame provides protection for young plants in winter.

CREATING A COLD FRAME

Cold frames are wooden boxes with glass roofs that give plants some protection against the elements. Although they do not offer frost-free conditions, you can use them to overwinter hardy young plants that may suffer in cold, wet weather and to house seedlings in late spring before the weather is warm enough to plant them outside. To make your own, first find a recycled window frame with the panes intact. Construct a box the same size as the frame, using the method for making a raised bed (see pp.90–91). Buy two heavy-duty stainless steel hinges and screw them to the window and the box on one of the longer edges, so that the frame sits neatly on the timbers when it is closed.

Plastic bottles on the top of bamboo canes will sway and rattle in the breeze, scaring birds away.

MAKING A WATER BUTT

Water butts made from old oak barrels offer an eco-friendly alternative to plastic models, which are not biodegradable and may split over time. You can buy barrel water butts or make your own from an old whisky, beer, or wine barrel, obtainable from specialist suppliers and online. Check that the barrel has a lid or solid top to prevent debris falling in and that it has not been used to store chemicals or treated with preservatives. Place the barrel on bricks so that a watering can will fit under the tap – you can add a wooden tap, available from barrel suppliers, if yours does not have one already. Site the barrel no more than 50cm (20in) from a drainpipe, then fit a rainwater diverter to the pipe, following the instructions on the kit. This will ensure the barrel does not overflow.

Barrel water butts can store large volumes of rainwater for use on plants.

WEAVING A PLANT SUPPORT

Young willow stems, known as withies, have been used for centuries to make woven structures such as plant supports. Ideal for runner and French beans, they also make beautiful rustic supports for climbing flowers, such as sweet peas and compact clematis. You can grow your own willow (*Salix alba*, *S. viminalis*, or *S. purpurea*) in the garden if you have moisture-retentive soil and a sunny spot, or buy withies via mail order.

1 To harvest withies, in early spring, use loppers or a pruning saw to cut all the stems from a young willow tree to one or two buds from the trunk at about head height. This technique, known as pollarding, encourages the plant to put on new straight growth for weaving. Repeat this process every two or three years.

2 Insert up to seven sturdy stems in the ground to create the uprights for the support, using a terracotta pot as a guide. Tie them at the top with the younger flexible willow stems or strong twine. Soaking stems in water for two or three days will increase their flexibility if they have dried out.

3 Weave the stems around the base a few times. Tie them in place with more flexible willow stems or strong twine. Repeat two or three times up the support to create a strong structure.

4 Install your plant supports in the garden wherever they are needed.

REDUCING YOUR GARDEN'S FUEL CONSUMPTION

From power tools to patio lights, the fuel used by modern garden equipment contributes to pollution levels and climate change – but even a brief look at how earlier generations maintained their gardens shows that much of today's energy consumption is not necessary.

Traditional low-tech methods can be employed to reduce your carbon footprint without affecting the beauty or productivity of your patch. Here are ways to cut your fuel needs while benefiting both your own health and that of the planet.

WORKING OUT IN THE GARDEN

Electric tools make it easier for us to keep the garden in good order, but the price that we pay is the environmental damage they cause. It is an interesting paradox that in the 21st century we may now expend less of our own energy on jobs in the garden but then go to a gym to exercise, when working outside has more benefits for our health and helps to reduce pollution too. Research shows that gardening increases muscle strength and bone mineral density, and lowers cholesterol levels and blood pressure, contributing to a healthier and longer life. Scientists have also discovered that exercising outside in a natural environment is more beneficial than exercising indoors, as it stimulates an enzyme called telomerase, which helps to prevent age-related illnesses. The answer is to swap power tools and machines for mechanical equipment such as push mowers and ratchet loppers that use no polluting fuels. Instead of reaching for a leaf blower, invest in a besom broom made from birch twigs, and when cleaning a patio or deck, scrub it with hot water and a stiff brush, rather than using a power washer.

TOP TIP REMEMBER NOT TO USE DETERGENTS WHEN CLEANING IN THE GARDEN – THEY WILL SEEP INTO PONDS AND WATERWAYS, HARMING WILDLIFE AND INCREASING POLLUTION.

Using a push mower is good exercise and also avoids the use of petrol or electricity.

A besom broom is an efficient and eco-friendly tool for sweeping up leaves.

TURN DOWN THE HEAT

The trend for heating outdoor seating areas has raised many concerns about the environmental damage that patio heaters and fire pits can cause. Exterior gas-powered heaters have a high carbon footprint, and while electric models that use infrared technology to heat your body rather than the air may be more energy efficient, they still have a cost. Burning coal or wet wood – banned in some countries – in a firepit is also polluting, producing particulates that affect our health and air quality. Even locally sourced, natural products, such as freshly cut branches from your tree, are not safe to use – one study showed that burning wet wood for an hour is more polluting than driving a diesel truck. The question is, do we really need to heat outdoor spaces at all? A warm blanket will keep you snug on cool evenings and is perhaps more effective than a heater. You can also surround a seating area with a hedge windbreak to create more shelter, thereby reducing the need for heating and fuel use.

Surrounding your seats with hedging will help to keep out the cold.

To prevent light pollution having an adverse effect on animals and plants, always switch off lights after use at night.

LIGHTING THE WAY

Most garden lights today are either LEDs, which use very little mains electricity, or solar-powered units that use none at all. However, while this is a positive step, they can still cause light pollution, which has a negative effect on wildlife. Studies have shown that LEDs often found in street lights emit wavelengths of blue light that affect animals – including humans – more than lights emitting from other parts of the spectrum. Light pollution disrupts normal animal behaviour; for example, blackbirds should only wake at dawn, but they have been heard singing all night in brightly lit cities. Other creatures affected by light pollution include bats, migratory birds, fish, and insects, and it can even have an impact on plants. Solar-powered units may be a better choice because they emit low light levels, but they too will have a detrimental effect if they shine all night, so remember to turn them off before going indoors.

TOP TIP AVOID CHEAP SOLAR-POWERED LIGHTS THAT BREAK EASILY. THEY ARE OFTEN MADE FROM PLASTIC AND A RANGE OF OTHER MATERIALS THAT MAKE THEM DIFFICULT TO RECYCLE, AND MOST GO TO LANDFILL AFTER A YEAR OR TWO.

DRY RIGHT

Making space in the garden to dry your clothes can have a significant impact on your energy needs. Almost 75 per cent of American and 50 per cent of European households own a tumbledryer, which has one of the highest fuel consumptions in the home, but you can dry your clothes outside for free. Doing so is also better for fabrics, since studies show that dryers break down the fibres in materials. Air-dried laundry smells better, too, and you can add to that sweet scent by growing some lavender or rosemary bushes under your clothesline.

Drying your washing on a clothesline uses no fuel and is better for the fabrics.

REDUCING WATER USE

Water is a precious resource, even in areas where there is relatively high rainfall. The water that pours from our taps has been processed, filtered, and made safe to drink, usually with the addition of chemicals such as chlorine, all of which entails an energy cost and environmental price tag. Try to minimize its use in the garden by preserving the rainwater that falls on your plot, choosing and positioning your plants carefully, particularly if you live in a drought-prone region, and only watering those plants that really need it.

Covering the soil with plants will prevent high evaporation rates.

Adding mulch to your soil makes it more moisture-retentive.

PRESERVING SOIL WATER

One of the most efficient ways to reduce our reliance on tap water is to trap rain that falls on the garden. In spring, after rain has soaked the soil, apply a 5–10cm (2–4in) layer of organic mulch, such as well-rotted animal manure or homemade compost, over the surface (see also pp.48–53). This will reduce evaporation and maintain soil moisture at deeper levels. As rainwater drains deep into the soil, it encourages plant roots to follow it down to where moisture is more consistent, which helps to protect plants from periods of drought when upper soil levels are dry.

SELECTING AND SITING YOUR PLANTS

To reduce your water needs, look for plants that will suit your local climate and garden conditions. Use drought-tolerant species in areas with low rainfall, in sunny spots on sandy soil, and at the tops of slopes. Good choices include lavender, rock roses (*Cistus*), sea holly (*Eryngium*), Russian sage (*Perovskia atriplicifolia*, syn. *Salvia yangii*), and lamb's ears (*Stachys byzantina*). Low-growing plants adapted to arid conditions, such as sedums, houseleeks (*Sempervivum*), and Cape daisies (*Osteospermum*), are ideal for containers that dry out quickly.

Where rainfall – or lack of it – is not predictable, which now includes many more areas due to climate change, drought-tolerant plants may not always be the answer. For example, if your soil is bone dry in summer but saturated in winter, lavender and other plants adapted to arid conditions will not survive the wetter months. Look instead for plants that can cope with a range of soil conditions and some shade, and plant them away from spots that receive the full glare of the summer sun. Good choices include buddleias, catmint (*Nepeta*), many hardy geraniums, and astrantias – and try others that thrive in part-shade to see which work well in your garden.

Lavender, salvias, and thymes are good choices for dry soils.

Sedums and houseleeks need very little water, even in summer.

WHAT TO WATER

Mature plants with extensive root systems should not need watering; if they do, you probably have the wrong plant in the wrong place (see pp.40–41). Even during long periods of drought in summer, most trees and shrubs should cope, assuming rain returns within a month or two. Established perennials that are adapted to your garden conditions should need no or very little extra irrigation and lawns that have dried out and turned yellow are also drought-resistant and will usually green up again after it has rained. This means that very few plants really need irrigating. Focus your watering can on seedlings, young plants with immature root systems and those you have just planted, annual crops, and plants in pots and containers. Bear in mind, too, that recent research has shown that many crops have higher levels of health-promoting phytochemicals if they are slightly water-stressed.

Target the water from your can on the soil above the root zone.

HOW TO WATER

Plants absorb water largely through their roots, so avoid splashing their leaves or flowers, which will have little effect. Use a can with a rose head or a hose on a gentle spray and target the soil above the root zone. Give your plants a long drink so that moisture drains down to the lower depths – small, frequent doses mean the water remains close to the surface where the roots are more vulnerable to drying out. Take care when watering plants in pots, where the moisture may run off the leaf canopy and on to the ground. Water two or three times a week, unless pots are small, when a daily dose may be needed. Apply water early in the morning or in the evening, when evaporation rates are at their lowest.

Pots of summer bedding will need to be watered every few days.

Water cabbages and other crops only during dry spells.

Established trees and shrubs suited to your climate and soil conditions will thrive without extra irrigation during dry spells.

NEED TO KNOW

Do not rush to water plants that wilt on hot days in summer. For many, this is a temporary state that allows the plant to preserve its water content by not releasing it through the stomata (pores in its leaves) as the plant would do normally. Wait until evening falls and the temperature cools, and only water then if the plant is still wilting.

COLLECTING AND REUSING WATER

Capturing and storing rainwater that would otherwise go down the drain is the perfect way to make the most of precious supplies. It also helps to prevent water running off the garden and into overloaded street drainage systems, where it can pick up pollutants and contaminate waterways. Rainwater collected in butts or buckets is ideal for irrigating plants because it is free from chlorine, which is added to tap water to make it safe to drink but is disliked by some ornamental plants and crops. In a few instances, household water can also be saved and used to irrigate plants, depending on the cleaning products you use.

Prevent water from being wasted by capturing it from a shed roof.

RECYCLING RAINWATER

To reduce waste and irrigate plants during dry spells, recycle as much rainwater as you can from the roofs and hard surfaces in your garden. Install water butts to the downpipes on your home, garage, shed, and greenhouse and use the rainwater on thirsty plants and crops (see p.99). Water from a butt is also ideal for topping up ponds and water features – tap water can disrupt a natural aquatic ecosystem. Also include a butt to capture rainwater in a front garden to prevent excess run-off from flowing into the street drainage systems, which can become overloaded and cause pollution during and after a storm (see pp.30–31).

CHOOSING A WATER BUTT

Plastic butts are relatively cheap, but they will obviously bring more of this non-biodegradable material into your garden. However, in a small space, a slimline plastic model may be your only option. To minimize the impact on the environment, opt for a good quality butt made from recycled plastic. You can also buy galvanized steel butts, which may be more eco-friendly but can heat up quickly in summer, causing some of the water to evaporate. Alternatively, opt for a recycled wine or whisky barrel made from oak and sustainable metals (see p.95). Any water butt should be sealed at the top or have a tight-fitting lid to prevent leaves and other debris falling in and polluting the water.

The easiest way to install a butt is with a rainwater diverter, which you attach to a downpipe. You can also choose to divert the overflow from a butt into other vessels so that no rainwater runs down into the drain when the butt is full – just make sure that the collection tanks are large enough to prevent water from spilling over the sides and flooding the garden.

Butts made from recycled plastic are inexpensive and practical.

NEED TO KNOW
- If you use your water butt often, the water should remain clean and clear. If it starts to smell, lift the lid and remove any organic matter and debris such as moss and leaves that may be causing the problem.
- Never use bleach to clean a water butt – it will kill your plants.
- If you use a cleaning product designed for water butts, check the label to ensure that it is safe for wildlife.
- Do not use water from a butt to fill bird baths. It may have picked up grime and bird droppings from the roof and gutters that could cause harm to birds.

OTHER WAYS TO CAPTURE WATER

If you have no suitable place to install a water butt, you can still collect some of the rainwater that falls on your garden. Place metal buckets or water storage tanks with open tops away from trees and they will soon fill up during a downpour. You can also use open vessels on a roof terrace or balcony to collect water for house plants, most of which prefer rainwater to that from a tap. In the garden, remember to fix wire mesh over the top of your containers to prevent small creatures and plant debris from falling into them.

Use the collected water before it becomes stagnant and starts to smell. If it is not needed immediately you can pour it into clean bottles and store them in a cool spot to keep the water fresh for a little longer.

Create a decorative feature with water-capturing bowls, but remember to cover them with mesh to prevent wildlife and debris falling in.

Water used for washing vegetables and salads can be saved for the garden.

USING GREY WATER

Any water that is used in the home, apart from water flushed down the toilet, is known as grey water. It includes water from the bath, shower, and bathroom and kitchen sinks, and while some is safe to use on the garden, its suitability depends on the products you use. Most dish-washing liquids contain detergents that contribute to pollution when they enter a waterway (see p.30). Detergents can also harm fish and other aquatic life in a pond – you may be surprised to find this warning hidden discreetly on the label of even so-called "gentle" products. Biodegradable soaps can have the same negative effect, especially if they contain petroleum, colours, and fragrances.

Water that contains products made from natural oils such as coconut oil, shea butter, olive oil, aloe vera, and sustainably sourced palm oil can be used sparingly on ornamental plants and crops, as long as it does not touch the edible parts – do not pour it on root vegetables and leafy crops such as lettuce leaves, for example. Use as soon as the water cools in a bath or sink; never store it, because the soap will soon attract pathogens that may cause diseases. Also, do not apply water that contains soap residues in areas where it will drain into a pond or water feature. Water in which you have scrubbed your veggies before cooking or serving is also safe to use on any plants and near water features – place a bowl in the sink to capture it.

MAKING A HOME FOR WILDLIFE

**Eco-gardens are havens for wildlife, offering homes to a
rich diversity of insects, birds, bats, frogs, toads, and small
mammals. Supporting the many species that are drawn to
our gardens is important conservation work, helping to
reverse the worrying declines in bees, butterflies, and other
creatures whose numbers are falling due to climate change
and loss of habitats. So take up the challenge and discover
how to create your very own nature reserve for all to enjoy.**

FEEDING GARDEN WILDLIFE

One of the best ways to encourage wildlife to visit your garden is to provide suitable food and water. While bird feeders and supplementary treats can help, plants play a much greater role in supporting the life in your garden. Planting areas with a diverse range of species will form the foundation of a complex food web that sustains the many creatures in your plot, from tiny beetles that live in the soil to foxes, birds, and bats.

THE GARDEN ECOSYSTEM

Natural ecosystems are sustained by a web of complex interactions between the plants and animals that are found in a particular habitat. The same is true of garden ecosystems, with the difference that conditions in adjacent gardens can vary enormously; for example, you may have a wildflower meadow, while your neighbour may have a planting of conifers. Not only are gardens profoundly influenced by human intervention, they are generally very small, so they have less resilience than larger natural ecosystems.

The highly simplified food web below shows how organisms might interact in a garden ecosystem: plants, both living and dead, provide food for herbivores and decomposers, which are in turn eaten by a range of predators. What the diagram highlights is that making a change to one aspect of the ecosystem – for example, by spraying insecticide to kill aphids – will very likely cause the entire system to adjust with consequences that may not be predictable or desirable.

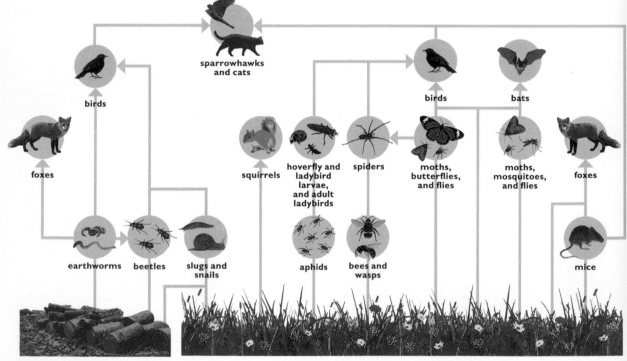

sparrowhawks and cats

birds

birds

bats

foxes

squirrels

hoverfly and ladybird larvae, and adult ladybirds

spiders

moths, butterflies, and flies

moths, mosquitoes, and flies

foxes

earthworms

beetles

slugs and snails

aphids

bees and wasps

mice

Dead plants

Live plants: green growth; nectar; pollen; fruits; nuts

This simplified garden food chain diagram shows the interdependence of many species of wildlife.

Providing a range of habitats will attract a wide variety of wildlife.

MAXIMIZING FOOD SOURCES

Clearly, plants are the key to feeding all of the wildlife in your garden and including as many different species as possible will help to create a haven for the broadest range of creatures. Trees are among the most important habitats. Their foliage provides food for the larvae of beetles, moths, and other insects, while the fruit and berries they produce feed birds, insects, and small mammals. Planting one or two trees in your garden, including a fruit-bearing species, will encourage more wildlife to move in (see pp.18–19).

Also plant a rich diversity of flowers for pollinators (see pp.126–133) and provide space for native species to flourish; some members of the local insect population will be adapted to feed on these in particular. A menu of shrubs and leafy species for different insect larvae to feast on will also bolster wildlife numbers. In addition, offer food sources throughout the year to prevent barren months when wildlife could suffer. A garden filled with summer bedding plants that offer minimal sources of nectar and pollen when in flower, and no food at all in early spring and winter, will be depleted of wildlife. Conversely, gardens with trees, hedges, and mature shrubs support the largest and most diverse wildlife populations.

LIVE AND LET DIE

You may not immediately think of fallen leaves as a food source for the wildlife in your garden, but these and other dead plant materials, such as roots and fallen branches, feed the decomposers, which form an essential part of the ecosystem. Decomposers include worms, beetles, and woodlice, upon which other predatory creatures depend, as well as bacteria and fungi. To feed them, allow autumn leaves to decompose naturally on the soil and place prunings in a quiet corner of the garden to rot down.

Dead plant material provides food for the decomposers in the garden.

The diet of adult wasps includes fruit, but they also feed insects to their larvae.

GARDEN PREDATORS

In many cases, the predators that eat the plant-eaters in your garden are your friends, offering a natural form of pest control. Birds eat insects, of course, but creatures that may not be among your favourite forms of wildlife, including spiders, centipedes, and even wasps, are equally important pest predators and they also provide food for other species further up the food chain. All play their part and should be encouraged, or at least tolerated, for their role in protecting your plants.

SUPPLEMENTING FOOD AND HABITATS

The garden food chain diagram shows that creating a garden rich in plants is the best way to feed wildlife, but there are times when a little extra help is needed. Offering supplementary food to birds, for example, and providing sites for insects to breed and hibernate, will contribute to their welfare. A water source is also important and a pond will create yet another thriving ecosystem to enrich your garden (see pp.28–29).

A log insect home will encourage more species to visit your garden.

FOOD FOR BEES AND OTHER POLLINATORS

There are about 20,000 species of bee worldwide, and all pollinate flowering and fruiting plants, including food crops. However, pesticide use, climate change, disease, and habitat loss have contributed to a dramatic fall in bee numbers and some species have become extinct. To help reverse this worrying trend, you can support bees and other important pollinators by growing a wide range of flowers and fruits.

Honey bees transfer pollen from plant to plant and also carry it back to their hives in pollen sacs behind their legs.

BEES IN CRISIS

The crash in bee populations is a worldwide phenomenon. Falls of up to 25 per cent in the numbers of native bees have been recorded in the USA in recent years, with similar declines in Europe. Bees of all kinds contribute to the pollination of 75 per cent of food crops, including staples such as apples, berries, and tomatoes, so any help gardeners can offer is an important investment in the future of these valuable insects. As well as planting flowers to feed them, banish pesticides and herbicides from your garden. Many studies have shown that pesticides cause bees to become weak and disorientated, so that they cannot locate food, while their ability to fight infections and produce eggs and sperm is also reduced.

Bees pollinate flowers as well as many important food crops.

Solitary bees do not fly long distances and need food close to their nests.

TYPES OF BEE

There are many different types of bee, and each has slightly different mouth parts that allow them to target different flowers. For example, some species of bumblebee have very long tongues that can reach deep into flowers such as those of lavender and honeysuckle to drink nectar and collect pollen, while the tongues of honeybees are much shorter, so they focus on flowers with more accessible sources of food.

The life cycle of bees determines when and where they need food supplies. Solitary species such as mason bees, which do not live in hives, spend the winter as

Open single flowers offer bees easy access to their pollen and nectar.

pupae, and because the adults do not fly long distances, they require food close to their nests when they emerge in spring (see p.124). Bumblebee colonies die in autumn and the old nests are abandoned, but the queens survive, hibernating through winter and appearing in spring to forage for food before making nests for their eggs, which hatch later in the summer. Honeybees overwinter in their hives, using their honey reserves to survive, but they may come out to feed on warm days. Bees' life cycles and warmer winters due to climate change means that bees can emerge when there is little food available for them, so provide nectar and pollen sources all year round.

FOOD SOURCES

Since different types of bee have a preference for different plants, try to include a wide diversity of species. For example, bumblebees love *Pulmonaria* in spring and lavender in summer, while honeybees are partial to spring blossom, sedums, and ivy for late-season nectar. The greater the diversity of plants in your garden, the more species of bee will pay a visit. You can even include bee-friendly plants, such as zinnias, cosmos, and snapdragons (*Antirrhinum majus*), in pots on your patio or balcony.

Select a variety of plants that will give a succession of flowers through the year. Good choices for winter include mahonia, with its sweetly scented yellow blooms; winter aconites (*Eranthis hyemalis*); snowdrops (*Galanthus*); and hazel trees (*Corylus*), which produce pollen-filled catkins from late winter. Follow with spring bulbs, such as crocus and grape hyacinths (*Muscari*), and blossoming trees. Also pack your garden with summer- and autumn-flowering plants that will provide a rich source of food when bees are most active (see pp.126–133).

Mahonia flowers feed pollinators if they fly on warm days in winter.

Grape hyacinths are easy-to-grow bulbs packed with bee food.

TOP TIP PROVIDE LONG GRASS AND UNDISTURBED LEAF LITTER TO ALLOW BUMBLEBEE QUEENS TO REST IN SAFETY BETWEEN FLIGHTS FOR FOOD IN EARLY SPRING. IF YOU FIND A BUMBLEBEE QUEEN ON THE PATIO OR OTHER HARD SURFACE AND SHE DOES NOT MOVE FOR A WHILE, GENTLY TRANSFER HER TO A MORE SHELTERED SPOT OR A BEE-FRIENDLY FLOWER.

Hoverflies resemble small wasps and are harmless pollinators.

OTHER IMPORTANT POLLINATORS

Bees may be the best known of the pollinating insects, but others also play an important role.

- Beetles were the first pollinators to evolve more than 160 million years ago, and they still contribute to pollination in our gardens today.
- Flies, wasps, and hoverflies, which do not sting but look like wasps to protect themselves from predators, are other key players, as are butterflies.
- Moths are useful too, pollinating night-scented plants that have evolved to attract them (for advice on what to plant to sustain these beautiful insects, see pp.108–109).

FEEDING BUTTERFLIES AND MOTHS

One of the best-loved of all insects, butterflies brighten the garden from spring to autumn. Night-flying moths may not occupy quite the same place in our affections, but they play an equally important role in the ecosystem. The larvae and adults of both provide food for birds, and are an important source of nutrients for chicks at nesting time. When these insects emerge from their pupae in spring, they also join the garden's army of pollinators. The numbers of butterflies and moths are plummeting worldwide, but we can help to boost falling populations by providing essential food and new habitats.

Nectar-rich daisy-like flowers attract many butterfly species.

FEEDING MOTHS AND BUTTERFLIES

Much media attention is given to the decline in the numbers of bees, but scientists believe that butterfly and moth populations are at even more risk. These insects react quickly to environmental changes, such as temperature rises and habitat loss, and their future is now in jeopardy. Scientists studying 28 butterfly species in the UK discovered that their numbers have fallen by almost 70 per cent since 1995. The situation is no better in the USA, where monarch butterfly numbers have dropped by 80 per cent since the mid-1990s. Meanwhile, more than 50 moth species in the UK became extinct during the 20th century and the picture is bleak worldwide for these night-flying pollinators.

REVERSING THE TRENDS

Butterflies and moths are important pollinators, and their decline is having a devastating effect on some plants and on birds, which eat both adults (especially moths) and their larvae. Their falling numbers have been attributed to intensive agricultural methods and the use of pesticides, but scientists point to habitat loss and air pollution as the main culprits. In towns and cities, paved-over gardens and reduced biodiversity are also cited as factors – problems that gardeners can easily remedy. Planting a wide range of species, and leaving an area for wild plants and flowers, will help to support butterfly and moth larvae. Some may munch on your ornamentals but in many cases they do no long-term harm, so simply use other plants to disguise any damage. Including nectar-rich flowers that adult butterflies and moths feed on (*see opposite*) will help to bolster numbers, too.

Butterflies use their long tube-like proboscis to reach the nectar.

PLANTING FOR BUTTERFLIES

You may find the numbers of butterflies increase in your garden if you plant native trees and shrubs, which supply food for their larvae. Grasses and weeds, such as nettles, are the favourite plants of some native butterfly species, so set aside an area in the garden for these too. Nectar sources are needed for the adults, and good supplies are particularly important in spring as they emerge and in autumn, when they need to build up energy reserves to survive hibernation. Good plant choices for spring include fruit tree blossom, alliums, and sweet rocket (*Hesperis matronalis*), while asters and sedums will provide late-season nectar. Butterflies like warmth, so choose sunny, sheltered areas for your plants. Select a wide variety and plant each type in groups to attract a range of butterfly species.

PLANTS FOR ADULT BUTTERFLIES
Allium (*Allium*) • Buddleia (*Buddleja*) • Cornflower (*Centaurea cyanus*) • Globe thistle (*Echinops ritro*) • Sea holly (*Eryngium*) • Sweet rocket (*Hesperis matronalis*) • Sedum (*Hylotelephium*) • Lavender (*Lavandula angustifolia*) • Honeysuckle (*Lonicera periclymenum*) • Mint (*Mentha*) • Catmint (*Nepeta*) • Phlox (*Phlox drummondii* or *P. paniculata*) • Sage (*Salvia officinalis*) • Aster (*Symphyotrichum*) • Lilac (*Syringa vulgaris*)

Adult moths and butterflies lay eggs on their caterpillars' favourite food plants.

Plant wild flowers such as cornflowers to attract butterflies to your garden.

Sea holly species (*Eryngium*) are favourite foods of butterflies.

PLANTING FOR MOTHS

Like butterfly larvae, many moth caterpillars eat the leaves of native trees and shrubs, while others prefer wild flowers or grasses. Research plants that are native to your locality and include some of them in your garden to lure in breeding insects. Most adults are active after sunset and attracted to pale-coloured, night-scented plants, many of which have evolved to attract these insects to pollinate their flowers.

PLANTS FOR MOTHS Buddleia (*Buddleja*) • Jasmine (*Jasminum officinale*) • Honeysuckle (*Lonicera periclymenum*) • Night-scented stocks (*Matthiola longipetala*) • Tobacco plant (*Nicotiana*) • Evening primrose (*Oenothera biennis*) • Petunia (*Petunia*)

Jasmine's sweet scent attracts night-flying moths to the flowers.

LIFE CYCLES EXPLAINED

Many moths and butterflies lay their eggs on the leaves of a host plant – usually a specific type which the caterpillars are adapted to eat. Other species drop their eggs into long grass during flight, while those whose eggs overwinter often lay them on a tree stem. The eggs then hatch into caterpillars, which spend most of their time feeding. Once a caterpillar attains its full size, it forms into a pupa, also known as a chrysalis. Inside the pupa, the caterpillar undergoes the final metamorphosis before emerging as a butterfly or moth.

PROVIDING FOOD FOR BIRDS

Gardens can provide the perfect habitat for many types of bird, both those that are in residence all year and migrants that may visit for a few months to breed. Offering food from a feeder is a great way to encourage birds into the garden, while including plants that provide them with their favourite foods, such as seeds and berries, is another way to keep them healthy during the autumn and winter months.

Waxwings eat mainly berries and are particularly fond of those from rowan trees (*Sorbus aucuparia*).

Thrushes and blackbirds enjoy a protein-rich diet of worms.

WHAT DO BIRDS EAT?

Birds are omnivores, eating insects, worms, and molluscs, as well as nuts, seeds, and berries. Grow plants such as trees and flowers that attract insects to provide them with a rich diet year after year. Lawns also harbour insects that birds enjoy, as well as worms that rise to the surface after rain, which make a good meal. Allowing some grass to grow longer and bloom will lure even more insect species for birds to eat.

Native trees and shrubs not only harbour insects that birds feed on, some also supply a feast of fruit, berries, and nuts. Other flowering plants, such as sunflowers, honesty (*Lunaria annua*), and asters, produce protein-rich seeds that help to sustain birds as the cold weather approaches. See also pp.134–139 for more planting suggestions.

PLANTING A BORDER FOR BIRDS

To maximize the number of birds visiting your garden, plant up a border in a sunny or partly shaded spot with seed- and fruit-bearing flowers and shrubs. Start by removing pernicious weeds, stones, and debris from the site in early autumn. Plant a few large shrubs, such as firethorn (*Pyracantha*), elder (*Sambucus nigra*), and guelder rose (*Viburnum opulus*) at the back of the border, checking their

heights and spreads to make sure that you have space for them to grow. Add plants that produce seeds that birds enjoy (*see pp.134–139*) in front of the shrubs, planting each species together in groups of three or more. Include early and later-flowering plants, such as tickseed (*Coreopsis*) with globe thistles (*Echinops*), to extend the seed supply. Water the border after planting and add a mulch of organic matter such as homemade compost over the soil surface. On sandy soils, wait until spring to add the mulch.

Firethorn cultivars provide a feast of red, orange, or yellow berries for birds through autumn and winter.

Position freestanding feeders in a safe, open area where birds cannot be ambushed by cats.

POSITIONING BIRD FEEDERS

A richly planted garden will offer a menu of delights for birds, but additional food will help to sustain them in winter and when they are breeding. Provide a range of feeding stations, such as bird tables and hanging feeders, to suit different species. Position tables in a quiet area where you can see the birds without disturbing them, and with good views all around so that they will not be ambushed by cats or other predators. A nearby look-out post, such as a tree branch, will allow the birds to check that it is safe to feed. Also include some metal hanging feeders in similar sites.

Some birds prefer to feed on the ground, so scatter food on a lawn or patio, or put out a ground-feeding tray. Give just enough each day so that surpluses do not attract vermin.

HOMEMADE BIRD FEEDER

You can make your own feeders by gently heating one part suet or lard with two parts bird seed and raisins in a saucepan. When the fat has just melted, pour the mix into a mould; you can also stir in some grated cheese as it cools. To incorporate a length of twine for hanging, use a mould with a hole in the bottom and insert the twine before adding the food mix. Suspend the feeders from a branch where birds will be safe from predators.

Feeders are easy to make from suet or lard, seeds, and raisins.

OFFERING CHOICE

A table set with seeds, soaked raisins, oats, apples chopped into small pieces, and mealworms will attract a variety of birds, but do not include whole peanuts, which may choke small species and chicks. Peanuts can be given in a feeder made from wire mesh. Check that the nuts you buy are guaranteed to be free from aflatoxin, a fungus that kills birds, and crush them slightly with rolling pin. Fill other hanging feeders with mixed seeds. Birds also enjoy fat balls, but do not buy them covered with plastic mesh, or if you do, remove it before setting them outside.

NEED TO KNOW
- Keep bird tables and surrounding areas free from droppings or mouldy food to prevent parasites and bacterial diseases.
- Clean your tables and feeders regularly with hot soapy water.
- Move feeding stations every month to prevent droppings accumulating underneath them.
- Wash your hands thoroughly after cleaning bird feeders.

Hanging feeders are popular with tits and woodpeckers.

SOWING A MEADOW

Sowing a wildflower meadow in your garden will attract wildlife of all sorts, such as bees, butterflies, and moths seeking nectar, and birds that eat the seeds. Your meadow need not be large – a small patch in a sunny or partly shaded area will provide space to plant some native grasses and flowers. Sow them in early autumn when the soil is warm and moist, and follow the tips here to ensure your plants establish successfully and bloom year after year.

In small gardens, sow a few wild flowers into existing beds.

Pictorial meadows, which include colourful cornfield annuals, will need to be sown year after year to maintain the effect.

WHAT IS A MEADOW?

A true wildflower meadow comprises perennial plants that appear each year. They include native grasses and flowers such as greater knapweed, ox-eye daisies, clover, bird's-foot trefoil (*Lotus corniculatus*), buttercups, meadow cranesbill (*Geranium pratense*), and bush vetch (*Vicia sepium*), among others. These meadows are quite subtle in colour and are usually dominated by a few flower species. They only require sowing once, and are mown each year in summer.

The images that many people associate with meadows are actually beds of cornfield annuals, sometimes referred to as "pictorial meadows". These are filled with grasses and colourful annual plants such as cornflowers (*Centaurea cyanus*), poppies (*Papaver rhoeas*), corncockles (*Agrostemma githago*), and corn marigolds (*Glebionis segetum*). The flowers provide a useful source of nectar for pollinators, but they will bloom just once and then disappear if you do not resow them annually or turn the soil in autumn to promote self-sowing. For the best of both worlds, you can sow a few of the more eye-catching flowers in pots and plant the seedlings into an established wildflower meadow each spring (see p.61). Alternatively, add a few annual cornfield plants to a mixed border.

SITING A MEADOW

Most meadow plants require a sunny, open area and poor soil, which will ensure that the grasses do not dominate the flowers. There are meadow mixes for shady sites too, but your choice of flower species may be more limited. Although large meadows have great impact, wildflowers will flourish on just a few square metres in small to medium-sized gardens too. When sowing from scratch, select an area away from mature trees and shrubs, which can inhibit seed germination.

Most soils are too rich for a flower meadow and you will need to strip 8–15cm (3–6in) from the top layer and sow on the infertile soil beneath (use the soil you have removed in pots or raised beds). Alternatively, sow a crop of mustard plants in the first year, which will use up many of the nutrients.

Wild flowers will thrive in a meadow sown on poor soil.

HOW TO SOW
A WILDFLOWER MEADOW

YOU WILL NEED Wildflower seeds suited to your conditions • Horticultural sand • Bamboo canes • Spring-tined rake • Garden rake

1 Remove some topsoil if necessary (*see opposite*), as well as weeds and large stones. Rake the soil to break up any large lumps. Leave the bed for two or three weeks, then weed it again. To prevent the seeds washing away, water the area before sowing. Use canes to mark out the area into square metres.

2 Mix your seed with dry horticultural sand to help spread it evenly. Weigh out roughly 5g of flower and grass seed per square metre (¼oz per square yard) of soil and mix it with the sand in batches in clean containers.

3 Sow each square with a batch of the seed and sand mix, using a spring-tined rake to lightly cover the seeds before moving on to the next section. Avoid walking on the areas you have sown.

4 With the back of a garden rake, gently press down the sown bed to ensure that there is good contact between the seed and soil, which will aid germination. Net over the bed to protect it from birds, which are very partial to flower seed.

MAINTENANCE

After about 6–8 weeks of growth in spring, mow the plants to 5cm (2in), and repeat every two months throughout the first summer. In subsequent years, cut the meadow annually in late summer once the flowers have set seed. After mowing, leave the meadow clippings on the surface for a few days to encourage self-seeding, then remove and compost them so that they do not rot down *in situ* and enrich the soil.

INCREASING YOUR GARDEN'S HABITATS

Gardens offer vital support to creatures whose homes have been lost due to habitat depletion. Creating areas for wildlife will help to protect species by providing them with safe havens, food, and good conditions for breeding. Boosting the numbers of birds, mammals, amphibians, and invertebrates in your garden helps increase the genetic diversity within each species, making their populations more resilient.

Viburnums will shelter butterflies during rain showers while also offering berries for birds in autumn.

CREATING NEW HABITATS

A garden typically includes a diversity of plants that support a range of wildlife species (see pp.104–105). Diversity is important, because the wider the choice of plants, the greater the number of creatures you will attract – and once they have taken up residence, if a plant that they feed on fails, others will offer an alternative source of nutrients.

Even in a small plot, you can increase habitats with imaginative planting. Pack your terrace or patio with flowering plants and shrubs in containers; clipped yew topiary in a large pot, for instance, will provide a nesting site and autumn berries for birds. Use shelves to make the most of vertical planting space and fix hanging baskets and pots to house walls, filling them with nectar-rich flowers and herbs loved by pollinators (see pp.126–133). Where possible, cover walls and fences with climbers that offer food and nesting sites – ivy is one of the best for wildlife, but you will need to keep it in check by clipping it back annually after any fledglings have departed.

A vertical container garden adds more diversity on a patio.

Diverse plant species support the widest range of garden wildlife.

Grow ivy in a pot to provide food for pollinators or on a fence for birds to nest in.

MAKING A BOG GARDEN

By creating an area of damp ground in a dry garden, you can introduce a new range of plants to increase biodiversity. For the most natural effect, make your bog garden near a pond or small pool (see also pp.28–29) and choose a partly shaded spot so that it does not dry out too quickly. Use rainwater from a butt, or tap water if you have no other source, to keep the soil damp in hot weather.

YOU WILL NEED String or chalk • Spade • Pond liner or recycled plastic sheet • Fork • Gravel or coarse grit • Rake • Well-rotted organic matter • Scissors • Bog plants

1 Mark out an area using string or chalk and dig out the soil to a depth of about 60cm (24in). Place the liner or plastic sheet in the hole and push it into the corners. Weigh down the edges at the top of the hole. Using a fork, pierce the liner at 60–90cm (2–3ft) intervals – you want the area to retain water but not become completely waterlogged.
2 To ensure that the drainage holes do not become blocked over time, cover the liner with a 8cm (3in) layer of gravel or coarse grit.
3 Fill the bog garden with the soil that you excavated, along with well-rotted organic matter. Using scissors, cut visible excess liner from around the edges. Add a few cans of water from a water butt and leave to drain.
4 Plant up and add a layer of organic matter over the surface.

PLANTS FOR WILDLIFE-FRIENDLY BOG GARDENS Gravel root (*Eupatorium purpureum*) • Meadowsweet (*Filipendula ulmaria*) • Water avens (*Geum rivale*) • *Hosta* species • Leopard plant (*Ligularia* 'The Rocket') • Gooseneck loosestrife (*Lysimachia clethroides*) • Purple loosestrife (*Lythrum salicaria*) • Bee's primrose (*Primula beesiana*) • Candelabra primula (*Primula japonica*)

A HERB WINDOW BOX

To increase biodiversity in your garden, fill a window box with a variety of pollen-rich plants. This box is planted with a range of herbs, including thyme, purple basil, and wild marjoram (*Origanum vulgare*), which will attract pollinating insects during the flowering season. Fill gaps with lettuces, and pick just a few leaves at a time to extend the harvest. When the herbs come into bloom, hoverflies, parastic wasps, and many species of bee that help to control plant pests will visit them for their pollen. In subsequent years, when the thyme and marjoram outgrow the box, add them to a vegetable patch or flowerbed, and renew the container with younger plants.

Thyme, marjoram, and purple basil jostle with lettuces in this windowbox.

PROVIDING WATER FOR WILDLIFE

A pond is a great addition to a wildlife garden (*see pp.28–29*), but if this is unsuitable for your space, there are plenty of other ways to supply water sources. Bird baths are easy to make and provide water for both drinking and bathing, or you can install a small water feature that thirsty garden creatures can use. Just remember to keep the water clean to prevent the spread of disease.

In cold weather, check bird baths regularly to make sure the water is not frozen over and inaccessible to birds.

Bees will quench their thirst in bird baths or puddles in the garden.

A UNIVERSAL NEED

All forms of wildlife need water to live. Insects such as bees require only a little fresh water to sustain them, and they can often be seen drinking from puddles or water droplets on plants. Birds and small mammals need a water source from which they can drink safely, away from predators such as cats and where there is no risk of falling in and drowning. A pond with sloping sides is ideal for wildlife, but in small spaces, you can use other features to satisfy thirsty garden creatures.

WATER FOR BIRDS

As well as providing birds with food, it is important to offer a clean source of water, especially when natural supplies are frozen in winter or have dried up during the summer. Birds need water for drinking and for bathing, which loosens dirt on their feathers and makes preening easier. You can install one or two bird baths in the tiniest of spaces and they are generally safe for gardens used by young children, especially if you elevate them on stands.

Whether you buy a bird bath or make one (*see opposite*), position it carefully. Birds can become preoccupied and vulnerable when bathing, so it is vital to make the bath as safe as possible. Place it where birds have a clear view all around, with shrubs or trees close by to provide cover and perches for preening after bathing. To prevent cats from creeping up on birds, position your baths near thorny species such as holly or pyracantha. Alternatively, place stems of prickly plants on the ground around other trees and shrubs where cats may hide.

Vulnerable young birds need a site that is safe from predators while they bathe, as flight is hampered by wet feathers.

An upturned dustbin lid secured in place on bricks makes an ideal vessel for sociable species to bathe together safely.

MAKING YOUR OWN BATH

A good bird bath should have shallow sloping sides, allowing birds to perch on the edge and drink safely. A water depth of 2.5–10cm (1–4in) is ideal. Also make sure that the surface is rough-textured, so that birds can grip it and avoid slipping into the water. To attract the largest number of birds, make your bath as big as possible – a flock of sparrows or starlings will soon use up the water in a small vessel.

To make a simple bird bath, place a stone in the middle of a plant pot saucer made from terracotta or other textured material and fill with water. Place it on an old garden chair or table, a sturdy upright log, or an open-topped timber bird table. For a larger bath, upturn an old plastic or metal dustbin lid and add some large pebbles or stones. Support the lid with bricks or soil so that it does not tip over. The stones in both features provide areas for small birds and insects to land safely without falling in the water.

NEED TO KNOW
Many birds and other garden creatures drown during the summer when their natural sources of water have dried up and they attempt to drink from open-topped water butts or steep-sided water features. Either cover these features or make them safe by placing a plank of wood or a branch over them to provide a perch from which birds and other creatures can drink without falling in.

SOURCES FOR SMALL CREATURES

Small bubble fountains fitted with a solar-powered pump can also provide a water source for birds, insects, and small garden creatures. A feature that spills water on to a pebbled surface is a good choice, and the movement helps to prevent the water from becoming stagnant. To top up the feature, use fresh water from a butt or tap water – the latter is the safest because water from a butt may have picked up dirt and bird droppings. Never use water that has an odour, as it may be carrying diseases.

After cleaning your bird bath, refill it with fresh tap water.

KEEPING WATER CLEAN

Clean your bird bath weekly, if possible, and change the water every day or two. To prevent algae, decaying leaves, and bird faeces from polluting the water and causing diseases, scrub the sides and bottom with a stiff brush and just-boiled water, taking care not to scald yourself in the process. Remove any stones and clean them too. Rinse everything well with fresh tap water. For stubborn dirt, you can use a non-toxic bird bath cleaner, but ensure you rinse the bath over a sink or drain afterwards. If the water becomes frozen in winter, pour a little hot water in to melt it.

Birds can drink safely from the shallow water in a pebble feature.

CREATING A POND ECOSYSTEM

A pond is a magnet for garden wildlife, from birds and amphibians, including frogs, toads, and newts, to aquatic insects such as pond skaters and water boatmen. A well-planned pond will attract large numbers of species, creating a small aquatic ecosystem within the wider garden environment. To maximize the habitats your pond offers, include a range of water depths and planting zones, each of which will provide a home for different types of wildlife. If your plot is not large enough for a pond, you can still enjoy some water plants and aquatic wildlife in a smaller feature, such as an oak-barrel pool or glazed pot.

PLANTING UP A POND

Research has shown that covering 50–75 per cent of the surface of a pond with plants makes the best home for wildlife. If you have a large pond, try to include a selection of non-invasive, submerged aquatics, such as water lilies and common water crowfoot (*Ranunculus aquatilis*), which will help to control pond weeds by mopping up nutrients and excluding light with their leaves. When choosing a water lily for a small pond, opt for dwarf

Plant a range of species at these approximate depths in your pond to maximize the habitats for wildlife.

varieties such as one of the *Nymphaea* 'Pygmaea' cultivars. If the sides of your pond are shelved (*see p.29*), use the shallower depths for marginal plants, which will support an array of wildlife. In other settings, you can support these plants in their baskets on clay bricks or flat stones, so that they are sitting at the correct water level for their needs.

To keep pond plants in check, it is usually best to plant them in unfertilized garden soil in aquatic pots, which are fine-meshed plastic containers. As an alternative, you can recycle black plastic plant pots by punching small holes into the sides with a skewer to allow water to enter, or use small hessian sacks.

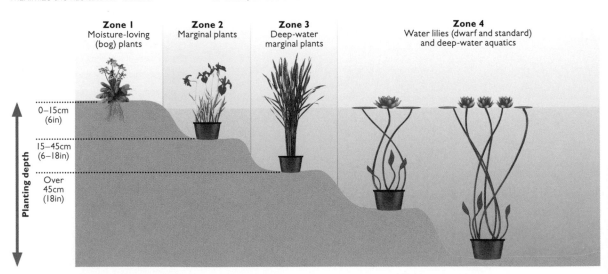

Zone 1 Moisture-loving (bog) plants

Zone 2 Marginal plants

Zone 3 Deep-water marginal plants

Zone 4 Water lilies (dwarf and standard) and deep-water aquatics

Planting depth

0–15cm (6in)

15–45cm (6–18in)

Over 45cm (18in)

WHO LIVES WHERE?

Once you provide a pond, wildlife will start to appear. Frogs and toads rapidly locate water sources, and dragon- and damselflies will pay a visit in summer. Taking a closer look into the water, you will notice many different types of insect, including water boatmen floating upside down just below the surface, pond skaters walking on the water with their spidery legs, and beetles hiding between the plant stems, as well as pond snails and the larvae of mayflies, dragonflies, and damselflies. It is best not to introduce fish if you want to attract local wildlife because they eat other creatures, such as tadpoles.

Scoop duckweed from the surface of your pond with a long-handled net.

CLEARING THE SURFACE

If your pond becomes covered in duck weed (small round leaves that form a film over the surface) and algae, you can scoop it out with a net, or twirl algae around a stick to remove it. Place the weeds close to the edge of the pond for a day or two so that any wildlife can crawl out and back into the water, then compost them. Never use chemical weedkillers in a pond.

Frogs often find a pond of their own accord, or friends and neighbours may be able to offer some tadpoles from their own gardens.

PLANTS FOR A POND

ZONE 1 Bog plants (see p.115)

ZONE 2 Flowering rush (*Butomus umbellatus*) • Marsh marigold (*Caltha palustris*) • Blue flag (*Iris versicolor*) • Water forget-me-not (*Myosotis scorpioides*) • Arrowhead (*Sagittaria sagittifolia*)

ZONE 3 Sweet flag (*Acorus calamus*) • Pickerel weed (*Pontederia cordata*) • Dwarf reed mace (*Typha minima*)

ZONE 4 Miniature water lilies (*Nymphaea odorata* var. *minor*; *N.* 'Pygmaea Helvola'; *N.* 'Pygmaea Rubra') • Larger water lilies (*Nymphaea* 'Froebelii'; *N.* 'Lemon Chiffon'; *N.* 'Marliacea Albida'; *N.* 'Rose Arey')

MAKE A MINIATURE POOL

If you do not have space for a full-sized pond, you can make a small feature in a watertight glazed pot or half oak barrel, available from specialist suppliers. Provide "stepping stones" by means of upturned pots and plant baskets to allow water creatures such as frogs and toads to get in and out easily, and to prevent other wildlife from drowning. Plant with pond plants, such as a miniature water lily, *Iris versicolor*, and zebra rush (*Schoenoplectus lacustris* subsp. *tabernaemontani* 'Zebrinus'). You will need to divide these in future years (see p.44) when they outgrow the space. Top up your feature regularly during warm or windy weather with water from a butt.

Waterlilies and rushes suit a small water feature for wildlife.

MAKING A HAVEN FOR BIRDS

Creating a home for birds is easy to achieve with a few shrubs and a tree or two. These plants will provide safe nesting sites as well as a food source, tempting many species to raise their young in your leafy haven. Installing nesting boxes will add to your garden's attractions, protecting brooding birds and their chicks from the elements and predators. Just make sure that any boxes are in a safe spot and securely fixed to a structure.

A wooden nesting box fixed to a shed and partially screened by greenery offers camouflage from predators.

Trees and large shrubs provide safe nesting sites while their twigs and leaves supply the building materials.

Nesting boxes come in a range of styles suited to particular species.

CREATING SAFE PLACES

Providing birds with places to perch, preen their feathers, and roost at night will help to attract many species to your garden. Birds look for safe places, such as trees and tall shrubs, from which they can survey the landscape for food and where predators cannot reach them. If you have space, include a fruit tree and an evergreen, such as yew (*Taxus baccata*), which offers shelter in winter as well as food.

From late winter and throughout spring, birds will also be searching for suitable places to build their nests. Small birds are attracted to hedges of prickly plants such as holly (*Ilex*), hawthorn (*Crataegus*) and blackthorn (*Prunus spinosa*). Many birds also appreciate evergreen climbers, especially ivy, which will create a leafy screen on a wall or fence that will keep chicks warm and well hidden. Other birds, such as swifts and swallows, may use the dry area under the eaves of your house to build their nests. Larger birds look for mature trees, where they can weave their sturdy nests out of harm's way – the amazing structures made by members of the crow family high up in the trees remain intact even when blasted by storms. Trees, shrubs, and grasses provide material for nest-building, while mosses, mud, and spiders' webs help birds to knit the structures together.

CHOOSING A NESTING BOX

While trees and shrubs can provide excellent nesting sites, man-made boxes offer vulnerable eggs and chicks extra protection from predators and the elements. You can buy nesting boxes or make your own, but remember that different birds require different styles. Some tend to choose those with a small hole at the entrance, while others are drawn to open-fronted boxes. To satisfy the needs of several species, fix a few different boxes around the garden. Buy your boxes from specialist sellers such as bird protection charities and wildlife organizations to make sure the designs and materials are suitable.

SITING A BOX

In autumn, fix most nesting boxes 2–4m (6–12ft) from the ground on a wall, fence, or tree trunk. Open-fronted boxes for robins and wrens need to be below 2m (6ft) but camouflaged by greenery. Use sturdy wire to attach a box to a tree, and add some cushioning material behind it to protect the trunk; do not use a nail, which may damage the tree. Make sure the entrance is not in full sun, which could make the nest too hot, and that it is sheltered from prevailing winds. Also tilt the box slightly forward so that rain runs off the roof, well clear of the entrance. Avoid placing two boxes close together and make sure that no plants or features near a box provide hiding places for predators such as cats.

Fix a bird box in part shade and away from prevailing winds.

ANNUAL CARE

From mid- to late autumn, lift the lid and remove the old nest from the box. Pour boiling water on to the box interior to kill any parasites and leave it to dry out completely before replacing the lid. Do not use insecticides. You can add a small handful of clean hay or wood shavings (not straw) to the box after cleaning, which may encourage small mammals to hibernate in it or birds to use it for roosting in winter.

In autumn, remove old nesting material, then clean out the box.

KEEPING YOUR DISTANCE

If you suspect that a nest or box in the garden has occupants, do not be tempted to get too close to have a look. This can frighten the parent birds and in the worst cases may cause them to abandon the nest. Once the eggs have hatched there will be extra activity to enjoy at a distance as you watch the parents ferrying food to their young ones. If you want to see the chicks as they develop, consider installing a box with an integral camera.

Venturing too close to chicks in a nest can frighten the adults, prompting them to abandon their young ones.

NEED TO KNOW

If you find a baby bird on the ground or on low branches calling for its parents, do not be tempted to rescue it. In most cases, the parents will be somewhere close by, encouraging the chick to learn how to feed and fly on its own. Human intervention may lead to the adult birds abandoning their chicks and they may then be reluctant to nest in your garden again. Only pick up a baby bird if it is in imminent danger, such as close to a road or a cat.

HOMES FOR MINIBEASTS

There are many tiny creatures in your garden that you may not be aware of until you uncover them beneath a fallen log or stone; others live under the surface, where they help to improve the soil structure. Frogs and toads also remain hidden for much of the time, only emerging from the shadows in quiet, damp areas when they are disturbed. These creatures, however small and seemingly insignificant, collectively make up the wider food web and ecosystem in your garden, and supporting them is as important as providing habitats for larger, perhaps more beautiful creatures such as birds.

Woodlice are important decomposers, helping to break down dead plants, which then enrich the soil with their nutrients.

GOING WILD

The best way to support this wildlife is to leave areas of the garden to go a little wild. Set aside a quiet area behind a shed or garage, allowing the grass to grow long and some weeds to flourish, which will provide the perfect habitat for amphibians, small reptiles, and many types of insect. You can also grow a few alpine strawberries in a partly shaded area to sustain small rodents and leave fallen fruits from your tree to languish on the ground, so that butterflies, beetles, and queen wasps can take their fill and stock up on essential calories before winter arrives.

Make a home for minibeasts with a range of native plants.

HIDDEN GARDEN LIFE

Insects and small invertebrates make up the largest groups of wildife in the garden, but most are hidden from view or only venture out after dark. However, look more closely and you will find a hidden world of vibrant life all around you. Turn over a fallen branch or clay pot and watch the woodlice and beetles scuttle off to find cover, or dig up a little soil to discover the insects and worms just below the surface. Mice and voles may also live in your plot, their presence only made known by nibbled fruit such as strawberries or a half-eaten flower bulb. All these creatures contribute to the ecosystem of your garden.

HABITATS IN DEAD WOOD

Many garden creatures live in or eat decaying or dead wood, and rotting logs and tree prunings provide them with an essential food source and habitat. Create a log pile in a damp corner where it will not dry out, and include a few upturned logs too (see p.124). The rotting wood will attract beetles, centipedes, and ladybirds, which collectively eat slugs, snails, and aphids, providing the perfect natural pest control. Hibernating frogs, toads, and newts are also attracted to damp wood piles and deep leaf litter which offer them shelter and a menu of mollusc- and insect-rich food. Other

Toads will help to keep the number of molluscs in your garden under control.

creatures that will take up residence in decaying logs include woodlice, ants, and other invertebrates that supplement the diets of garden birds.

ACCOMMODATING SMALL MAMMALS

As well as the tiny insect life in the garden, you can play host to slightly larger animals too. Wood mice and voles are generally nocturnal, although you may see them towards the end of the day in summer. They may eat some of your ornamental or crop seeds, but making them welcome in the garden can be a positive step, as they also eat weed seeds and plant pests and are in

turn an important food source for foxes, owls, and other birds of prey. Growing plants that they like in your wild area will deter them from eating your prized flowers or peas and beans, while hedges, shrubs, and trees offer them cover and nesting sites.

Shrews, which look like mice with long, pointed noses, may also make their home in your garden if it mirrors their natural habitats of woodland and long grasses. These creatures are a gardener's friends, since they have a voracious appetite for insect pests and slugs and snails.

THE SECRET LIFE OF STAG BEETLES

Among the largest members of the beetle family, stag beetles derive their name from the male's fierce-looking jaws (mandibles), which resemble the antlers of a stag. These beetles have become endangered in Europe due to habitat loss and experts believe that many species in the US are also in decline, but you can help to support numbers by making a home for them in your garden. They lay their eggs underground and the larvae feed on dead and decaying wood, so to provide them with food, leave old tree stumps in situ, and create log piles on the soil. Stag beetles remain at the larval stage for up to seven years, depending on the weather, and then build cocoons in the soil, where they pupate and metamorphose into adults. When the adults emerge, they feed on plant sap and fallen fruits. They mate in summer, then die before winter – you may spot a male stag beetle sunning himself to gather strength before flying off to find a mate.

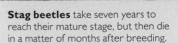

Stag beetles take seven years to reach their mature stage, but then die in a matter of months after breeding.

Wood mice may eat a few pea and bean seeds, but they provide valuable food for owls and other birds of prey.

MAKING INSECT HOTELS

You can buy bee and insect hotels to provide the minibeasts in your garden with living quarters, but it is easy to make them at home. Construct a wood pile and recycle wine boxes and terracotta pots into insect accommodation by filling them with materials collected from the garden. These minibeast residences will provide insects of all sorts with safe places in which to take cover and hibernate in winter.

A few hollow stems in a terracotta pot are enough to make a home for insects.

ASSEMBLING THE PERFECT WOODPILE

Woodpiles provide an ideal habitat for many types of insect (see *p.122*), and they are easy to make. Branches cut from ash, oak, beech, and birch measuring at least 10cm (4in) in diameter, with the bark still attached, will support the most species. Wood from other native trees, including heritage fruit trees, is also suitable. Place the large branches directly on the ground in dappled shade – not dense shade, which is too cold for most species. Choose a spot that is not too close to live plants, since decaying wood can harbour diseases. Insert smaller twigs and stems between the branches. If you have space, also include a few upright logs by partially burying stems of different widths to a depth of 45–50cm (18–20in).

If you do not have a large tree in your own garden that requires some pruning, seek out local tree surgeons and ask them if they are able to supply you with logs. Do not collect fallen branches from natural woodlands, where they will be forming part of the local ecosystem.

A woodpile in a quiet corner of the garden will house many beneficial insects.

CREATING HOUSES FOR SOLITARY BEES

Unlike honeybees and bumblebees, solitary bees do not live in colonies and most build individual nests in hollow stems and holes in the ground. The female solitary mason bee, for example, lays each of her eggs on a ball of pollen in a stem or hole, then constructs a partition wall with mud and repeats the process until the space is almost full; she usually leaves the last cell empty. She then seals the end with mud. When the eggs hatch, the larvae eat the pollen. After a few weeks each larva creates a pupa, in which they metamorphize into an adult bee and hibernate for the winter, emerging in spring to mate and start the cycle again. After hibernation, the adults live for just 10–12 weeks.

Studies show that a single mason bee pollinates 120 times more flowers than a honeybee. Create homes in your garden for these and other solitary bee species using a range of hollow stems that have cavities 2–10mm (1/16–3/8in) in diameter. Bamboo canes and plants such as buddleja, Himalayan honeysuckle (*Leycesteria formosa*), purple top (*Verbena bonariensis*), and brambles are ideal. Pack the stems into a terracotta pot and secure it to a fence, wall, or tree in a sunny, sheltered spot, ensuring that the entrance is tilted down towards the ground to prevent rain from entering.

MAKING A HOTEL FROM A WINE BOX

This homemade multipurpose bug hotel will attract a variety of insects, such as hibernating lacewings and ladybirds, as well as nesting solitary bees and wasps.

YOU WILL NEED Hollow stems, moss, dried leaves, pine cones, and other plant material • Wooden wine box • Paint (optional) • Hammer • Stainless steel nails • Picture hook • Recycled wood or slate tile for the roof (optional) • Drill with drill bits of different sizes • Short lengths of branch • Small terracotta pots • Terracotta pieces • Shavings from untreated wood

1 Collect together a variety of natural plant materials to fill the hotel. Remove the front sliding section from the wooden wine box. If you choose to paint the outside, use a non-toxic, solvent-free odourless paint that will not harm wildlife.

2 Using a hammer and nail, fix the picture hook to the top of the box. You can also make a small pitched roof from recycled wood to protect it from rain if you have good DIY skills, or drill a hole in a piece of slate tile and nail it to the top so that it overhangs the edges.

3 Drill several holes with varying diameters of 2–10mm (¹⁄₁₆–³⁄₈in) in the ends of three sections of tree branches. Select a couple of small terracotta pots that will fit into the box and fill these with broken pieces of terracotta, wood shavings, and leaves.

4 Fill the insect hotel with the hollow stems, terracotta pots, pieces of drilled wood, moss, leaves, and pine cones. Hang the hotel on a screw or nail fixed to a sheltered, sunny wall or fence, siting it at least 1m (3ft) above the ground.

NEED TO KNOW

• At the end of summer, keep your solitary bee homes healthy by removing any stems that have enclosed entrances from the previous year – they will probably contain dead bees.

• To prevent a build-up of fungus, mould, mites, and other pests and parasites, replace drilled branches with new ones every two years once the young bees have emerged.

• When buying ready-made bee and insect homes, make sure they are fit for purpose by purchasing them from a trustworthy supplier, such as a conservation charity.

PLANTS FOR POLLINATORS

Bring your garden to life with the sights and sounds of pollinating insects, such as bees, butterflies, and hoverflies. These beneficial bugs not only enable your plants to set seed and produce fruits, they also help to establish a healthy eco-system in your garden. The larvae of hoverflies and other pollinators contribute to the control of plant pests such as aphids, while insects in turn provide food for birds. Insecticides and diseases have reduced the numbers of bees and other pollinators worldwide, so including food for them will also help to redress this decline.

HOLLYHOCK *ALCEA ROSEA*

HEIGHT AND SPREAD 2 × 0.6m (6 × 2ft)
SOIL Well-drained
HARDY Fully hardy
SUN ☼

This towering cottage-garden staple is an imposing sight, with stout stems featuring hairy, lobed leaves, followed by large, cup-shaped flowers. Loved by bees and other pollinators, it is a biennial or short-lived perennial that comes in a range of colours from purple and red to pink, yellow, and white. Plant it at the back of a flower border and stake stems in exposed sites. If the leaves fall victim to rust, use other plants to hide the lower stems; this disease does not seem to reduce the hollyhock's ability to flower.

In summer, the tall spires of large, cup-shaped flowers make an impressive display.

LADY'S LEEK *ALLIUM CERNUUM*

HEIGHT AND SPREAD 50 × 20cm (20 × 8in)
SOIL Well-drained
HARDY Fully hardy
SUN ☼

Dainty pink flowers adorn this allium in summer, while slim green leaves form a skirt at the base. The small bulbs can be squeezed between other pollinator-friendly plants, maximizing wildlife benefits in a small plot. It is trouble-free if grown on well-drained soil. Plant bulbs in autumn for flowers the following year. It will then reappear annually and form a clump over time. It may also self-seed, providing plants elsewhere in the garden and creating a naturalistic effect. Remove unwanted seedlings in spring.

Fountains of bell-shaped flowers create an elegant feature in the summer garden.

BUTTERFLY FLOWER *ASCLEPIAS TUBEROSA*

HEIGHT AND SPREAD 90 × 45cm (36 × 18in)
SOIL Well-drained
HARDY Hardy to -10°C (14°F)
SUN ☼

As its common name suggests, this colourful nectar-rich perennial is a magnet for butterflies, but it is also loved by other pollinating insects. The clusters of small, bright orange-red or golden flowers appear on tall, sturdy stems above slim green leaves and sustain wildlife from midsummer to late autumn. They are followed by unusual seedheads, which are covered with cotton wool-like fibres. Grow this plant in a flowerbed or mixed border, or use it to stand proud above the grasses in a meadow.

The butterfly flower's colourful blooms stand out against the bright green leaves.

BUTTERFLY BUSH *BUDDLEJA DAVIDII*

HEIGHT AND SPREAD up to 3 × 5m (10 × 15ft)
SOIL Well-drained
HARDY Fully hardy
SUN ☼ ☀

A common sight on road verges and railway sidings, the butterfly bush has much to offer the wildlife in your garden, attracting many types of butterfly. Bees and other pollinators are also drawn to its long cone-shaped fragrant flowerheads, which appear in summer and come in a wide range of colours, from dark purple and lilac to bright pink and white. Use it along a garden boundary or at the back of a mixed border, and encourage an abundance of blooms by cutting the stems back hard in early spring.

NANHO PURPLE is a compact cultivar with fragrant flowers borne in dense panicles.

HEATHER *CALLUNA VULGARIS*

HEIGHT AND SPREAD 20 × 35cm (8 × 14in)
SOIL Well-drained acid soil/ericaceous compost
HARDY Fully hardy
SUN ☼

Moors covered with heather buzz with bees when the scented blooms appear from midsummer to autumn, and you can replicate this effect in your garden if you have the acid soil this plant demands (see p.41). The spikes of urn-shaped pink or pale purple flowers appear above small, narrow, evergreen leaves; the many cultivars widen the colour choice to include dark pink, red, and white forms. Plant it as ground cover and, in spring, use shears to cut back the flowered shoots to 2.5cm (1in) above the old growth.

'Alicla' bears very long-lasting white flowers from summer through to autumn.

CUCKOO FLOWER *CARDAMINE PRATENSIS*

HEIGHT AND SPREAD 40 × 30cm (16 × 12in)
SOIL Moist but well-drained
HARDY Fully hardy
SUN ☼ ☀

A beautiful perennial that often colonizes cool, shady, waterside spots in the wild, the cuckoo flower produces a rosette of leaves divided into rounded leaflets and clusters of pale pink or white flowers on upright stems in late spring or early summer. Plant it in drifts beneath trees in damp areas, in wet meadows, on the shady sides of a pond or in a bog garden (see p.115). Sow seed in pots indoors in autumn or spring, or outside *in situ* when the weather is warmer and seed is more likely to germinate.

The delicate blooms of cuckoo flower attract many pollinators to cool, shady spots.

CORNELIAN CHERRY *CORNUS MAS*

HEIGHT AND SPREAD up to 4 × 4m (12 × 12ft)
SOIL Well-drained/moist but well-drained
HARDY Fully hardy
SUN ☼ ☀

Brighten up the winter garden with this deciduous shrub or small tree which features clusters of tiny, bright yellow flowers on bare stems as the season comes to a close, offering early-flying pollinators a welcome source of food. The green oval leaves follow in spring and turn purple in autumn, after the edible red cherry-like fruits have formed. This easy-going plant is happy in most sites and soils and makes a good backdrop to borders planted with spring bulbs and flowers that bloom later in the year.

Bright yellow flowers on bare stems stand out in the winter and early spring garden.

COSMOS _COSMOS BIPINNATUS_

HEIGHT AND SPREAD 90 × 60cm (3 × 2ft)
SOIL Well-drained/moist but well-drained
HARDY Does not tolerate frost
SUN ☼

Few annuals are easier to grow than cosmos, and its cheerful daisy-like flowers can be used throughout the garden as well as in containers on a patio. The ferny green leaves make a beautiful textural contrast to the flowers, which bloom from summer to the frosts in autumn, if deadheaded regularly. Cultivars come in shades of red, pink, yellow, or white – select single or semi-double flowers for pollinators. Sow seed indoors in spring and plant out after the frosts in the middle of a mixed border.

'Psyche White' produces a continuous display of flowers from summer to autumn.

SPRING CROCUS _CROCUS VERNUS_

HEIGHT AND SPREAD 10 × 5cm (4 × 2in)
SOIL Well-drained
HARDY Fully hardy
SUN ☼

These bulbous perennials are the harbingers of spring, their cup-shaped flowers appearing at the beginning of the season when pollinators are just emerging from their winter hibernation. The blooms come in a wide range of colours, including purple, pink, yellow, and white, with contrasting orange stamens which deliver the pollen. Some are striped or patterned. Plant the corms _en masse_ in autumn in a lawn, at the front of a border, or in a rock garden. You can also grow crocus in shallow pots of gritty compost.

'Pickwick' is an early-flowering cultivar that has pretty purple flowers with white veining.

DAHLIA (OPEN-FLOWERED) _DAHLIA_

HEIGHT AND SPREAD 1 × 0.45m (3¼ × 1½ft)
SOIL Well-drained/moist but well-drained
HARDY Hardy to -5°C (23°F)
SUN ☼

Single- or peony-flowered dahlias make striking garden plants. These tuberous perennials are loved by many pollinators, but when you are choosing dahlias from catalogues or online, check that you can see the yellow stamens as some cultivars are so multi-petalled that pollinators cannot access them. Use dahlias in flower beds and borders, or in large pots on a patio. Plant tubers in early spring in pots indoors and set outside when the risk of frost has passed or plant in the soil later in spring.

'Bishop of Landaff' has dark foliage and produces its bright red flowers until the first frosts.

COMMON FOXGLOVE _DIGITALIS PURPUREA_

HEIGHT AND SPREAD 1.5 × 0.5m (5 × 1¾ft)
SOIL Well-drained/moist but well-drained
HARDY Fully hardy
SUN ☼ ☼

The common foxglove is a biennial, which means it produces leaves in its first year of growth and flowers the next year. Some plants may flower for a few more years, and given the right conditions they will self-seed. These stately plants produce a rosette of hairy, oval leaves and, in summer, tall spires of rosy-purple tubular flowers with spots inside. Buy plants or sow seed in pots in late spring and plant out at the edge of a tree canopy among other woodlanders, such as ferns and hellebores.

The spotted petals of foxglove blooms entice bees to stores of nectar and pollen inside.

PURPLE CONEFLOWER *ECHINACEA PURPUREA*

HEIGHT AND SPREAD up to 1 × 0.5m (3 × 1¾ft)
SOIL Well-drained
HARDY Hardy to -15°C (5°F)
SUN ☼ ☼

Loved by butterflies and bees, the perennial purple coneflower produces slim green leaves and sturdy stems topped by daisy-like pink flowers with slightly reflexed petals. Blooming from early summer to early autumn, it provides a rich store of pollen and nectar and birds enjoy the seedheads that follow. Dark pink and white cultivars are available, too. Grow it in bold drifts in a mixed border or among grasses and leave the seedheads through winter; cut back in spring to make way for new growth.

The coneflower gains its name from the shape of its prettily reflexed petals.

WINTER ACONITE *ERANTHIS HYEMALIS*

HEIGHT AND SPREAD 10 × 20cm (4 × 8in)
SOIL Moist but well-drained
HARDY Fully hardy
SUN ☼ ☼

The golden flowers of the winter aconite are guaranteed to add a cheerful note to the garden when little else is in bloom. The rich green ferny foliage appears with the cup-shaped bright yellow flowers, each bloom edged with a ruff of leaves. They produce a store of pollen for many weeks from mid- to late winter. Plant this perennial among other winter flowers such as snowdrops and crocuses, close to trees or in beds and borders – ideally in soil that will remain reasonably moist in summer.

Winter aconite's golden flowers and green foliage are an asset in the cold months.

SNAKE'S HEAD FRITILLARY

FRITILLARIA MELEAGRIS

HEIGHT AND SPREAD 30 × 10cm (12 × 4in)
SOIL Well-drained/moist but well-drained
HARDY Hardy to -15°C (5°F)
SUN ☼ ☼

The common name of this fritillary refers to the pattern on its nodding, bell-shaped purple flowers, which resembles snakeskin. It produces an underskirt of small, lance-shaped, greyish-green leaves, but it is the elegant flowers, loved by bees, which are the star attraction. Plant the fragile bulbs in autumn in an area that does not dry out completely in summer. If bulbs do not succeed, buy plants in flower in the spring.

This fritillary is well suited to a planting in a patch of rough grass or a meadow.

SPOTTED CRANESBILL *GERANIUM MACULATUM*

HEIGHT AND SPREAD 70 × 50cm (28 × 20in)
SOIL Well-drained/moist but well-drained
HARDY Fully hardy
SUN ☼ ☼

An ideal plant for beginners, this easy-going perennial produces large clumps of lobed green foliage that makes a textural carpeting ground cover. In late spring and early summer, it produces abundant pollen-rich pale mauve to pale pink flowers with a white eye, held on slim stems above the leaves. Cut back after flowering to promote new leaf growth and encourage a second flush of blooms in early autumn. Use this cranesbill at the front or middle of a mixed border, or beside trees.

The bright flowers of the spotted cranesbill stand out against the attractive foliage.

SNEEZEWEED *HELENIUM*

HEIGHT AND SPREAD 1 × 0.5m (3 × 1¾ft)
SOIL Moist but well-drained
HARDY Fully hardy
SUN ☼

Heleniums add a splash of vibrant colour to beds and borders. These perennial plants have simple green leaves and tall stems topped with red, orange, or yellow flowers from midsummer to early autumn, although some bloom slightly earlier. Add bold swathes to borders for maximum impact, or smaller clumps in beds where space is at a premium. Tall varieties will need stakes. The flowers are rich in nectar and pollen; leave the seedheads on plants to feed the birds and provide a home for hibernating insects.

'Sahin's Early Flowerer' is an early-blooming cultivar that continues to flower into autumn.

ENGLISH LAVENDER *LAVANDULA ANGUSTIFOLIA*

HEIGHT AND SPREAD 60 × 75cm (24 × 30in)
SOIL Well-drained
HARDY Hardy to -15°C (5°F)
SUN ☼

English lavender is a magnet for bees and hoverflies when the scented blue-purple flowers appear in summer. The blooms persist for several weeks above the aromatic grey-green, evergreen foliage. Grow it as a low hedge within a garden or to edge a flower bed. To keep plants neat and compact, lightly prune the flower spikes after blooming and again in spring, taking them down to new growth. This plant will withstand low winter temperatures on free-draining soil but will rot in cold, wet conditions.

The flowers of lavender provide a beautiful haze of colour in the summer garden.

POACHED EGG PLANT *LIMNANTHES DOUGLASII*

HEIGHT AND SPREAD 15 × 15cm (6 × 6in)
SOIL Well-drained/moist but well-drained
HARDY Hardy to -15°C (5°F)
SUN ☼

This easy-to-grow hardy annual produces a carpet of fern-like, bright yellow-green foliage and, from summer to autumn, abundant lightly scented, cup-shaped white flowers with yellow centres. They attract bees and hoverflies, the latter and their larvae offering natural aphid control. Sow seed in spring at the front of a border or in a rock or gravel garden, or sow in pots in early autumn, overwintering the young plants in a frost-free place. Given sunshine and the right soil, it will self-seed freely.

The bright yellow and white flowers of poached egg plant attract many types of pollinator.

HONEYSUCKLE *LONICERA PERICLYMENUM*

HEIGHT AND SPREAD 7 × 2m (23 × 6ft)
SOIL Moist but well-drained
HARDY Fully hardy
SUN ☼ ☼

Prized for its scented flowers, which are more intensely fragrant in the evening, this deciduous climber attracts a wide range of pollinators, including moths. The spidery pink and white or yellow flowers appear from late spring to early summer. In hot summers, they are followed by edible small red fruits. Use honeysuckle to cover a fence or wall or to grow through a shrub or tree. It may produce a second flush of blooms later in summer, particularly when stems are cut back after the first flowering.

Scented flowers amid fresh green leaves cover honeysuckle's twining stems in summer.

BERGAMOT *MONARDA DIDYMA*

HEIGHT AND SPREAD 90 × 45cm (36 × 18in)
SOIL Moist but well-drained/moist
HARDY Fully hardy
SUN ☀ ☀

Also known as bee balm, this decorative garden perennial produces aromatic, narrow green leaves and red, purple, or pink pollen-rich spidery flowers over many weeks from summer to early autumn. Choose from the many colourful cultivars, but ensure the soil is reasonably moist to prevent powdery mildew from developing or select a disease-resistant variety. Grow in groups in the middle of a flowerbed or border, alongside clumps of other pollinator-friendly perennials and shrubs.

'Cambridge Scarlet' bears bright flowers from midsummer until early autumn.

GRAPE HYACINTH *MUSCARI ARMENIACUM*

HEIGHT AND SPREAD 15 × 5cm (6 × 2in)
SOIL Well-drained/moist but well-drained
HARDY Fully hardy
SUN ☀ ☀

The green, grass-like foliage of the grape hyacinth appears in late winter and is joined in spring by spikes of tiny mid-blue flowers, loved by bees. Very easy to grow, this little bulb will spread to form clumps and will self-seed to create a naturalistic effect. Grow it close to trees, at the front of a border or in a wildlife or gravel garden; it also makes a colourful feature in patio pots. The top growth dies down over the summer. Cultivars offer a variety of colour choices, including sky blue, pink, and white flowers.

Spreading clumps of grape hyacinth offer easy spring colour and food for pollinators.

SICILIAN HONEY GARLIC
NECTAROSCORDUM SICULUM

HEIGHT AND SPREAD 120 × 25cm (48 × 10in)
SOIL Well-drained/moist but well-drained
HARDY Hardy to -15°C (5°F)
SUN ☀ ☀

A cousin of allium, this plant produces linear garlic-scented foliage and, from late spring to early summer, tall stems topped with a pretty fountain of cream, pink, and green-tinted bell-shaped flowers that are loved by butterflies and bees. They are followed by seed pods that form little castle-like turrets. The bulbs, best planted in groups in autumn, can be squeezed between perennials in small gardens. The stems snap easily, so choose a sheltered spot.

Intriguing flowers and dramatic seedheads earn this plant a spot in any garden.

COMMON POPPY *PAPAVER RHOEAS*

HEIGHT AND SPREAD 75 × 30cm (30 × 12in)
SOIL Well-drained
HARDY Fully hardy
SUN ☀

Historically, this vibrant wild poppy was a common sight in agricultural fields and it gained its place as a symbol of remembrance when its dormant seeds burst into bloom on the battlefields during World War I. It bears its characteristic bright red blooms with black eyes on wiry stems above fern-like green leaves. Sow the seed in autumn or spring on poor, well-drained soil, or as part of a meadow mix with grasses and other wild flowers – you will need to sow it annually to ensure a display every year.

'Mother of Pearl' is a pretty mix of single and double flowers in a range of colours.

FIDDLENECK *PHACELIA TANACETIFOLIA*

HEIGHT AND SPREAD 90 × 45cm (36 × 18in)
SOIL Well-drained
HARDY Hardy to -10°C (14°F)
SUN ☼ ☼

Also known as lacy phacelia and purple tansy, this tall hardy annual produces fern-like leaves and spikes of small pale blue or lavender-blue bell-shaped flowers over many months in summer. Loved by bees and other pollinators, it makes a good filler for cottage or prairie schemes, or grow it with grasses in a wildflower meadow. Sow *in situ* in early autumn, or in spring in colder areas. It can also be used as a green manure (see *p.55*), and will self-seed widely; remove unwanted seedlings in spring.

With its blue flowers on tall stems, fiddleneck offers wildlife and visual benefits over summer.

TURKISH SAGE *PHLOMIS RUSSELIANA*

HEIGHT AND SPREAD 90 × 50cm (36 × 20in)
SOIL Well-drained/moist but well-drained
HARDY Fully hardy
SUN ☼

The Turkish sage is a perennial which in mild areas can retain its large, rough-textured, grey-green leaves over winter. In early summer, hooded, soft yellow flowers packed with pollen appear along its tall, stout stems, accompanied by lance-shaped leaves. They are followed by bead-like seedheads that persist over winter and offer a home for hibernating insects. Grow it in bold clumps in the middle of a flowerbed or border, or in drifts in a prairie-style scheme. It thrives in sun, but will tolerate a little light shade.

Turkish sage's whorls of soft yellow flowers appear for a few weeks over the summer.

COWSLIP *PRIMULA VERIS*

HEIGHT AND SPREAD 25 × 25cm (10 × 10in)
SOIL Moist but well-drained/moist
HARDY Hardy to -15°C (5°F)
SUN ☼ ☼

Once established in conditions it enjoys, this wild form of cowslip self-seeds, brightening up the spring garden with its rosette of textured, paddle-shaped green leaves and upright stems of nodding, bell-shaped, yellow flowers. Sweetly scented, they are attractive to bees. Plant this little perennial in wildflower meadows and at the edge of woodland areas, or in borders in a wildlife garden. The seeds need low temperatures to germinate; sow them *in situ* in autumn, or buy young plants in early spring.

A classic wildflower plant, cowslip adds sunny colour to the garden in spring.

FLOWERING CURRANT *RIBES SANGUINEUM*

HEIGHT AND SPREAD 3 × 2.5m (10 × 8ft)
SOIL Well-drained/moist but well-drained
HARDY Fully hardy
SUN ☼

The flowering currant is a deciduous shrub that bursts into life in mid-spring with drooping clusters of small pink flowers, followed by blue-black berries. Cultivars offer plants with pollen-rich blooms in various shades of pink, white, and crimson. The rough-textured, lobed leaves unfurl after the flowers have opened. Use this currant at the back of an informal or wildlife border or along a boundary fence or wall. Stems can be pruned back after flowering in late spring to keep plants in shape.

'Pulborough Scarlet' is a popular cultivar with clusters of deep crimson flowers.

CONEFLOWER *RUDBECKIA*

HEIGHT AND SPREAD up to 2 × 1m (6 × 3ft)
SOIL Moist but well-drained
HARDY Fully hardy
SUN ☼ ☼

Rudbeckia species include annuals and perennials such as the tall *R. laciniata* and more compact *R. fulgida*. They have small green leaves and upright stems topped by large, pollen-rich daisy flowers with yellow or orange petals and a central green or dark brown conical disk. These appear from late summer to autumn, providing a feast for pollinators before winter sets in. Plant in bold groups in the middle or towards the back of a border and leave the seedheads on in winter to provide a home for hibernating insects.

R. fulgida var. deamii is compact, free-flowering, and ideal for the front of a border.

BALKAN CLARY *SALVIA NEMOROSA*

HEIGHT AND SPREAD up to 50 × 30cm (20 × 12in)
SOIL Moist but well-drained
HARDY Fully hardy
SUN ☼

This long-flowering perennial produces upright stems of greyish-green aromatic foliage. In summer and early autumn, spikes of tiny violet-blue flowers draw in a range of pollinators, including bees and butterflies. To prolong the flowering period and availability of insect food stores, remove the flowering stems as the blooms start to fade and the plant will soon throw up more. Pink cultivars are also available. Plant Balkan clary in groups alongside other pollinator plants towards the front of a border.

The cultivar 'Ostfriesland' is a small form, reaching just 45cm (18in), with bright blooms.

RED CAMPION *SILENE DIOICA*

HEIGHT AND SPREAD 80 × 45cm (32 × 18in)
SOIL Well-drained/moist but well-drained
HARDY Fully hardy
SUN ☼

Frequently seen growing wild in woodland glades, this short-lived perennial features clumps of downy, oval, green leaves and tall stems of small rose-pink flowers. They appear in late spring, enticing butterflies, hoverflies, and bees, and are followed on female plants by cup-shaped seedheads. When grown on poor soils, these dainty plants will self-seed freely to create a colourful naturalistic effect. Use red campion in a perennial wildflower meadow or close to trees and shrubs in a wildlife or informal garden.

The simple open flowers of red campion make a pretty addition to an informal garden.

PURPLE TOP *VERBENA BONARIENSIS*

HEIGHT AND SPREAD 2 × 0.5m (6 × 1¾ft)
SOIL Well-drained
HARDY Hardy to -10°C (14°F)
SUN ☼

Tall, wand-like stems studded with small, rough-textured leaves and branched clusters of small, purple, scented flowers, rich in nectar and pollen, have made this perennial a favourite with wildlife gardeners. Blooming from summer to autumn, it attracts a host of butterflies and moths as well as bees; the seedheads offer food for birds too. Use it *en masse* in a flowerbed or border, or grow it in a gravel garden, where it will self-seed widely. It needs free-draining soil and may rot over winter on heavy clay.

Heads of tiny purple flowers appear to almost float on the slender stems of this plant.

PLANTS FOR BIRDS AND OTHER GARDEN WILDLIFE

Many plants will help to sustain wildlife in your garden. Seeds, berries, fruits, and nuts are important food sources for birds and other small creatures, while trees provide nesting and roosting sites for birds. Flowers attract insects that are in turn a vital source of protein for young birds and mammals.

YARROW *ACHILLEA MILLEFOLIUM*

HEIGHT AND SPREAD 60 × 60cm (24 × 24in)
SOIL Well-drained/moist but well-drained
HARDY Fully hardy
SUN ☼

The flat white flowerheads that appear for many weeks in summer on the slender stems of this dainty perennial are rich in nectar, attracting insects that birds feed on. Later in the year, the plant produces seedheads that birds enjoy. The feathery foliage adds to the plant's charms and it works well in a wildflower meadow along with grasses that birds use to create their nests in spring. You can buy cultivars with pink, red, or orange flowers that add a colourful note to sunny areas of the garden.

The flowerheads of yarrow make a meal for the insects that birds feast on in summer.

JAPANESE BARBERRY *BERBERIS THUNBERGII*

HEIGHT AND SPREAD 1.5 × 1.5m (5 × 5ft)
SOIL Well-drained/moist but well-drained
HARDY Fully hardy
SUN ☼ ☼

Barberry is a deciduous shrub with prickly stems covered in small oval leaves that turn red and orange in autumn. In spring, small birds choose it as a safe nesting site, where they can be protected by its dense, spiny growth. The small yellow flowers that attract insects appear in spring, followed in autumn by small red berries that provide food for birds into the winter. There are many forms to choose from, including *Berberis thunbergii* f. *atropurpurea* 'Golden Ring', with purple foliage.

'Golden Ring' has purple foliage that turns red in autumn, with accompanying red berries.

SILVER BIRCH *BETULA PENDULA*

HEIGHT AND SPREAD up to 25 × 10m (82 × 33ft)
SOIL Well-drained
HARDY Fully hardy
SUN ☼ ☼

Popular for its white trunk and graceful cascading stems, silver birch is ideal for a medium-sized garden, while cultivars such as 'Youngii' offer an option for smaller spaces. A deciduous tree, it attracts huge numbers of insects that feed on it, which in turn provide food for birds, while the seeds offer an additional feast for them in autumn. Woodpeckers also make their homes in the trunks of mature trees. The triangular green leaves make a dazzling sight when they turn buttery yellow in autumn.

The silver birch is an airy presence in the garden and sustains many insects and birds.

JAPANESE QUINCE *CHAENOMELES SPECIOSA*

HEIGHT AND SPREAD 1.5 × 2.5m (5 × 8ft)
SOIL Well-drained/moist but well-drained
HARDY Fully hardy
SUN ☼ ☼

This deciduous or semi-evergreen shrub, often trained along a fence or wall, produces a dense network of thorny stems that provide nesting birds and other wildlife with a protective barrier against predators. The spring flowers attract insects that are then preyed upon by birds – cultivars with open, single blooms are the best choices for wildlife and they are available in a range of colours. The fruit that follows softens after a frost and is eaten by wildlife once it has fallen to the ground.

'Cardinalis' is a smaller cultivar that bears single scarlet flowers followed by aromatic green fruit in autumn.

GOLDEN CLEMATIS *CLEMATIS TANGUTICA*

HEIGHT AND SPREAD up to 5 × 3m (15 × 10ft)
SOIL Moist but well-drained
HARDY Fully hardy
SUN ☼ ☼

An easy-going deciduous climber, the golden clematis will weave through trees and shrubs or over a boundary wall, given support. It offers cover for birds during the summer and autumn, when it also bears its lantern-shaped golden flowers. These are followed by attractive silky seedheads from autumn that persist on the plant until the next spring, when birds use them to line their nests. For a good supply, cut one-third of the stems to the ground in early spring and prune the rest a few weeks later.

The seedheads of *Clematis tangutica* are both decorative and useful to nesting birds.

TICKSEED *COREOPSIS VERTICILLATA*

HEIGHT AND SPREAD 50 × 50cm (20 × 20in)
SOIL Well-drained/moist but well-drained
HARDY Hardy to -15°C (5°F)
SUN ☼ ☼

Combining a mound of ferny green foliage and small golden or pale yellow flowers over many months in summer, tickseed draws a range of birds including goldfinches into the garden to feast on the seeds that follow the blooms. The flowers also attract bumblebees and butterflies, so there is little to dislike about this versatile perennial plant. Easy to grow from seed in spring, the plants may even flower in the first year after sowing. Use tickseed to edge a flowerbed or plant it in a large pot.

The pretty yellow flowers of tickseed produce copious seeds for birds later in the year.

COMMON HAWTHORN *CRATAEGUS MONOGYNA*

HEIGHT AND SPREAD up to 8 × 8m (26 × 26ft)
SOIL Well-drained/moist but well-drained
HARDY Fully hardy
SUN ☼ ☼

One of the best trees for wildlife, the common hawthorn supports many insect species that provide a staple diet for beetles and birds such as wrens and blue tits. The tangle of dense, prickly branches also offer a nesting site and home for small hibernating creatures, including toads and wood mice. The white flowers appear in late spring after the small green lobed foliage has unfurled; these are followed in autumn by red fruits (haws), which are enjoyed by an even wider range of birds.

Hawthorn's clusters of white flowers are followed by berries that birds enjoy.

MEZEREON *DAPHNE MEZEREUM*

HEIGHT AND SPREAD 1.5 × 1m (5 × 3ft)
SOIL Well-drained
HARDY Fully hardy
SUN ☼ ☀

All parts of this deciduous shrub are poisonous to humans, but it has many benefits for the birds and the garden. Its highly fragrant, purplish-red flowers, which appear in late winter or early spring, attract early-flying pollinators. A few birds, including greenfinches, are also drawn to the clusters of round red berries that form in autumn. Its spreading, open habit makes a good addition to an informal or woodland garden, and the early flowers offer much-needed spring colour.

The flowers of mezereon are a boon for pollinators in late winter and early spring.

GLOBE THISTLE *ECHINOPS RITRO*

HEIGHT AND SPREAD 90 × 45cm (36 × 18in)
SOIL Well-drained
HARDY Fully hardy
SUN ☼ ☀

This perennial's spiny foliage and tall, sturdy stems topped with spherical, thistle-like blue flowers in summer make a dramatic impact in the garden. The blooms offer invaluable stores of pollen for bees, butterflies, and hoverflies and in autumn they form seedheads which birds will soon strip. Plant globe thistles in the middle or towards the back of a border. They will self-seed given the right conditions, so watch out for young plants in spring and move them to areas where they are needed.

Globe thistle produces dramatic blooms and then a feast of seedheads for birds.

BEECH *FAGUS SYLVATICA*

HEIGHT AND SPREAD up to 12 × 8m (40 × 26ft)
SOIL Well-drained/moist but well-drained
HARDY Fully hardy
SUN ☼ ☀

If allowed to reach maturity, this deciduous tree is suitable only for large gardens; it is usually treated as a hedging plant or clipped annually to keep its size in check. Its yellow-green young leaves turn green in summer and then russet-brown in autumn, lasting over winter on young stems. Birds nest in its dense growth and in autumn they enjoy the beech nuts, enclosed in bristly cases. Small mammals also benefit from the nuts. Prune in late winter or early spring before nesting begins.

Beech nuts in autumn are an attractive feature and offer a valuable food source for birds.

IVY *HEDERA HELIX*

HEIGHT AND SPREAD 2 × 2m (6 × 6ft)
SOIL Well-drained
HARDY Fully hardy
SUN ☼

Ivy is often maligned for its rampant growth, but it is highly valuable for wildlife. The dense evergreen growth is ideal for nesting and offers protection for hibernating insects, which provide birds with a rich food supply. Ivy flowers in the autumn, when few other nectar sources are available to insects, and the berries that follow are also food for birds. Choose a well-behaved cultivar, grow it on a fence or a wall with sound pointing, and keep it in check by cutting growth back in late winter or early spring.

'Heise' is a medium-sized cultivar with dense grey-green foliage that shelters birds.

SUNFLOWER *HELIANTHUS ANNUUS*

HEIGHT AND SPREAD up to 100 × 50cm (36 × 20in)
SOIL Well-drained
HARDY Hardy to -5°C (23°F)
SUN ☼

Easy to grow from seed, the annual sunflower is loved by wildlife and gardeners alike. Honeybees, bumblebees, and hoverflies are drawn to the nectar-rich yellow flowers in summer, while doves, finches, and other birds flock to the plants when the large seedheads form in autumn. Sow the seeds indoors in biodegradable pots in spring (see pp.42–43) and plant outside after the frosts. There is a wide range of cultivars in various sizes, some of them with amber or orange flowers.

The giant seedheads and flowers of the annual sunflower attract birds and pollinators.

CANDYTUFT *IBERIS SEMPERVIRENS*

HEIGHT AND SPREAD 30 × 60cm (12 × 24in)
SOIL Well-drained/moist but well-drained
HARDY Hardy to -15°C (5°F)
SUN ☼

Candytuft's white flowers resemble a lace tablecloth when they cover the small evergreen leaves in spring and early summer. This subshrub attracts pollinators such as bees and butterflies as well as snails, slugs, and caterpillars that provide food for birds. It can be used as low-growing groundcover to edge a path or flowerbed, or in a trough on a patio. Trim lightly after flowering to maintain the shape and allow stems time to regrow before the winter, when they will provide a refuge for insects.

Clusters of white flowers and leafy growth harbour insects that sustain birds.

HOLLY *ILEX × MESERVEAE*

HEIGHT AND SPREAD 3 × 2.5m (10 × 8ft)
SOIL Well-drained/moist but well-drained
HARDY Fully hardy
SUN ☼ ☼

Providing birds with shelter for nesting, this compact holly with bluish-green leaves is a good choice for eco gardeners with small spaces. It supports many types of insect which provide food for birds, while the bright red winter berries offer another valuable source in autumn and winter. For berries, you will need to plant a female clone, such as Blue Princess, and a male pollinating partner. In areas that experience freezing winds in winter, plant it in a sheltered position.

The bright red berries of Blue Princess stand out against the blue-tinged leaves.

ANNUAL HONESTY *LUNARIA ANNUA*

HEIGHT AND SPREAD 90 × 50cm (36 × 20in)
SOIL Moist but well-drained
HARDY Fully hardy
SUN ☼ ☼

Honesty is a hardy annual or biennial that produces large clusters of purple flowers in late spring and early summer above toothed, heart-shaped leaves. It attracts caterpillars that birds eat, and the purple, nectar-rich summer flowers are followed by flat, round, silvery pods containing seed that bullfinches and other birds will devour. Sow seed in spring and plant outside when temperatures rise. It flowers in the first or second year after sowing, and you can collect your own seed to replenish your supplies.

The leaves of honesty are liked by caterpillars, which in turn provide food for birds.

TUPELO *NYSSA SYLVATICA*

HEIGHT AND SPREAD up to 15 × 6m (50 × 20ft)
SOIL Moist but well-drained acid soil
HARDY Fully hardy
SUN ☼ ☀

Also known as black gum, this graceful tree is suitable only for medium-sized to large gardens, but is well worth growing if you have the space. Its branches offer shelter and nesting sites, while thrushes, waxwings, and other birds relish the small purple autumn fruits. Woodpeckers will also make a home in mature trees. The narrow oval leaves create a dazzling display of red, purple, yellow, and orange shades in autumn before falling. Check that your soil is acidic (see p.41) before buying.

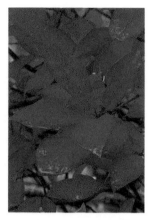
In autumn, tupelo trees put on a spectacular display of colours before the leaves fall.

WILD CHERRY *PRUNUS AVIUM*

HEIGHT AND SPREAD up to 12 × 8m (40 × 26ft)
SOIL Well-drained/moist but well-drained
HARDY Fully hardy
SUN ☼

The wild cherry is a beautiful tree, ideal for a medium-sized garden. The foliage supports many types of caterpillar, which in turn offer a food supply for birds, and in late spring the clusters of white flowers provide an early source of nectar and pollen for bees. Small, red-purple cherries appear in summer and are eaten by birds such as thrushes as well as small mammals. In autumn, the green oval leaves turn orange and red before falling and revealing the tree's skeleton of shiny reddish-brown stems.

'Sunburst' is a self-fertile cultivar with sweet, dark red fruits in midsummer.

FIRETHORN *PYRACANTHA*

HEIGHT AND SPREAD 2.5 × 2.5m (8 × 8ft)
SOIL Well-drained/moist but well-drained
HARDY Fully hardy
SUN ☼ ☀

The common name of this evergreen shrub describes both its thorny stems and fiery autumn berries, both of which offer benefits to birds and other garden creatures. The fierce thorns provide a protective defence for nesting birds and hibernating creatures, but take care to plant firethorn where it will not cause you harm, such as trained against a boundary wall. The pollen-rich spring flowers are followed in autumn and winter by a bird banquet of red, yellow, or orange berries.

The berries of firethorn light up this shrub in autumn and provide food for birds.

RED-LEAVED ROSE *ROSA GLAUCA*

HEIGHT AND SPREAD 2 × 2m (6 × 6ft)
SOIL Well-drained/moist but well-drained
HARDY Fully hardy
SUN ☼

Many old-fashioned shrub roses are disease-resistant and provide a rich food store for a range of garden wildlife. *Rosa glauca* is one of the most beautiful, with its arching, almost thornless, reddish stems covered with purple-flushed, greyish-green foliage. Clear pink flowers with pale centres appear briefly in early summer and are followed by orange-brown hips, which attract many types of bird. Tolerant of poor soil, this old rose makes a great addition to an informal garden.

The single flowers of *Rosa glauca* attract birds which prey on the insects that visit them.

BLACKBERRY *RUBUS FRUTICOSUS*

HEIGHT AND SPREAD up to 2.5 × 6m (8 × 20ft)
SOIL Well-drained/moist but well-drained
HARDY Fully hardy
SUN ☼ ☼

Wild blackberries make excellent habitats for birds, but in gardens they quickly smother everything in their path. The better-behaved cultivars are more suitable, and while some have thorny stems that offer protective nesting sites, the thornless types are easier to manage. All produce nectar-rich flowers that will attract pollinators, followed by the edible fruits which you can share with the birds from late summer to early autumn. Cut the fruited stems back to the ground in spring.

Blackberries provide a rich feast for wildlife from late summer to early autumn.

ELDER *SAMBUCUS NIGRA*

HEIGHT AND SPREAD 6 × 4m (20 × 12ft)
SOIL Moist but well-drained
HARDY Fully hardy
SUN ☼ ☼

Ideal for an informal garden, this large deciduous shrub or small tree provides birds and other creatures with a range of benefits. Its branches, clad with large, divided, green leaves, offer roosting and nesting sites, while the small black berries that follow the pollen-rich spring flowers deliver a rich source of nutrients in autumn; thrushes and blackbirds are especially partial to them. You can use both the flowers and berries to make drinks – cook them first, as both can cause discomfort if eaten raw.

Elderberries form in large clusters in autumn and are a favourite snack for birds.

NEW ENGLAND ASTER *SYMPHYOTRICHUM NOVAE-ANGLIAE*

HEIGHT AND SPREAD up to 1.5 × 1m (5 × 3ft)
SOIL Moist but well-drained
HARDY Fully hardy
SUN ☼ ☼

Plant this beautiful perennial for its daisy-like flowers, which appear from late summer to late autumn, depending on the cultivar. Available in shades of white, blue, pink, and purple, they bring a range of insects into the garden, including bees and butterflies, while the seeds that follow offer food for birds such as finches, helping to sustain them as winter approaches. Aster cultivars vary greatly in size, so look for those that will fit your space and plant them in groups for a colourful late-season display.

'Barr's Pink' produces its bright pink flowers with yellow centres in early autumn.

YEW *TAXUS BACCATA*

HEIGHT AND SPREAD up to 12 × 8m (40 × 26ft)
SOIL Well-drained
HARDY Fully hardy
SUN ☼ ☼

An evergreen tree with dark green needle-like leaves, yew can be clipped into a hedge, topiary or simply a smaller plant if you lack space for a mature specimen. It provides food and shelter for birds, together with other garden creatures which nest in its dense growth and feast on the small red berries. These appear on female plants in autumn if there is a male plant close by. Plant yew along a boundary or in a border as a backdrop to more colourful plants. All parts of it are poisonous.

Yew produces decorative red berries in autumn amid the dark green leaves.

INDEX

Bold text indicates a main entry for the subject.

A

Acer campestre **34**
Achillea millefolium **134**
aggregates 87
air quality 17, 24
Alcea rosea 42, 43, **126**
allium 21, 43, **72**, 109
Allium cernuum **126**
Amelanchier laevis **34**
Anemone hupehensis **72**
Antirrhinum majus **72**, 107
aphids 66, 68, 69, 71, 123
apple tree 17, **36**, 58
arrowhead 119
Asclepias tuberosa **126**
Aster × frikartii **73**
Astilbe 33
astrantia 98

B

Balkan clary **133**
barronwort 21
basil 66, 115
beans and peas 57, 69, 82, 83, 90, 95, 123
beech **136**
bees **106–07**, 115, 116
    food sources 107
    insect hotels 124, 125
    population crisis 106
Berberis thunbergii 25, **134**
bergamot **131**
bergenia 21, **73**
berries, edible 59
Betula nigra 25, **34**
Betula pendula 13, 25, **134**
bindweed 48, 55, 65
biosecurity 47
birch
    river birch 25, **34**
    silver birch 13, 25, **134**
bird cherry 25
birds **110–11**, 114
    bird bath 116–17
    bird feeder 111
    in food chain 104
    nesting and roosting **120–21**
    and pest control 66, 68
    scarer 94

birds and wildlife plants 134–39
    beech **136**
    blackberry 64, **139**
    candytuft **137**
    clematis 44, **74**, 95, **135**
    elder 25, 110, **139**
    firethorn 25, 110, **138**
    globe thistle 109, 110, **136**
    hawthorn 16, 24, 25, 35, 120, **135**
    holly 25, **36**, 116, 120, **137**
    honesty 43, 110, **137**
    ivy 9, 24, 25, 107, 114, 120, **136**
    Japanese barberry 25, **134**
    Japanese quince **135**
    mezereon **136**
    New England aster **139**
    rose, red-leaved **138**
    silver birch 13, 25, **134**
    sunflower 43, 69, 110, **137**
    tickseed 110, **135**
    tupelo **138**
    wild cherry **138**
    yarrow **134**
    yew 24, 25, 28, 114, 120, **139**
    see also wildlife, feeding
bird's-foot trefoil 60, 112
black walnut 17
black-eyed Susan 33
blackberry 64, **139**
blackthorn 121
bluebell 20, 21, 74
bog garden 13
Boston ivy 44
bramble 48, 55, 64, 65, 124
Buddleja 98, 109, 124, **127**
bugle 21, 32, 33
bush vetch 112
buttercup 60, 112
butterflies and moths 64, 107, **108–09**, 114
butterfly bush 98, 109, 124, **127**
butterfly flower **126**

C

cabbage family 57, 69, 99
    see also vegetable growing
Calendula 22, 69, **73**, 83, 90
Campanula poscharskyana 23, **73**
candytuft **137**
Cape daisy 98
carbon capture 16, 17, 20, 27
Cardamine pratensis **127**
catmint **76**, 98, 109
Centaurea montana **74**, 112
Chaenomeles 44, **135**
chamomile 32
Chinese anemone **72**

clematis 44, **74**, 95, **135**
climbing hydrangea 44
clover 55, 60, 61, 112
cold frame **94**
comfrey or nettle tea 63
companion plants 69
compost choices **50–53**, 61
    green manure, planting 55
    homemade 50, 51, **52–53**
    leafmould 51
    peat-free 50
    and planting preparation 49
    vegetable growing 55
    weed control 65
    see also feeding plants; mulching
coneflower 9, 33, **133**
    purple **129**
Coreopsis 110, **135**
corn marigold 112
corncockle 61, 112
Cornelian cherry **127**
cornflower **74**, 112
Corylus **35**, 44, 107
cosmos 67, 83, 107, **128**
couch grass 65
cow parsley 21
cowslip **132**
cranesbill 21, 75, 112
    spotted **129**
creeping jenny 32
crocus 107, **128**
crop rotation 57, 70
cuckoo flower **127**
cuttings 45
cypress 25

D

dahlia 66, 69, 83, **128**
dandelion 55, 60, 64, 65
daphne 21, 44, 76
Daphne mezereum **136**
dead nettle 20, 21
decking options 84, 85
Digitalis purpurea 21, **128**
disease defence **70–71**
dog's tooth violets 21
downy mildew 71

E

earwigs 69
Echinacea purpurea **129**
Echinops 109, 110, **136**
eggshells as pots 82
elder 25, 110, **139**
Eranthis hyemalis 107, **129**
eulalia 33
Euonymus 21, 25, 33, **35**
evening primrose 109

F

feeding plants **62–63**
    see also compost choices
fern 21, 22, 66, **74**
fiddleneck **132**
field maple **34**
firethorn 25, 110, **138**
flood defences **30–33**
    mini rain garden 33
    parking spaces and pathways 32
    rainwater and runoffs 30, 100, 101
    storm water collection 33
    trees 31
flowering currant 24, 25, **132**
foam flower 21
forget-me-not 32, 119
foxglove 21, **128**
French marigold 69
frogs, toads and newts 66, 67, 123
fruit growing 9, **58–59**
    organic certification 58
    seed harvesting 42
fuchsia 21, **74**
fuel consumption **96–97**
    solar-powered units 97, 117
Fuji cherry **37**

G

Galium odoratum 21, **75**
garden equipment and tools 10, **92–95**
garden furniture **86–87**
    see also landscape materials
garden ponds **26–29**
    banned plants 119
    as carbon sinks 27
    duckweed removal 119
    ecosystem **118–19**
    marginal plants 28
    miniature pool 119
    pond pests 67
    siting 28
    wildlife ponds 13, **28–29**
geranium, hardy 33, 98
Geranium sanguineum see cranesbill
germination tips 43
global warming, and trees 8
globe thistle 109, 110, **136**
grape hyacinth 107, **131**
gravel root 115
greening hard surfaces **22–23**
grey mould 71
ground cover 64
ground elder 65
groundsel 65, 70
guelder rose 21, 110

H

hawthorn 16, 24, 25, 35, 120, **135**
hazel **35**, 44, 107
heather **127**
Hedera helix 9, 24, 25, 107, 114, 120, **136**
hedge planting 25
Helenium **130**
Helianthus annuus 43, 69, 110, **137**
hellebore 21, **75**
heuchera 22
hoeing, and weeds 65
holly 25, **36**, 116, 120, **137**
hollyhock 42, 43, **126**
honesty 43, 110, **137**
honey fungus 71
honeysuckle 22, 109, 124, **130**
hornbeam 24, 25
horsetail 65
hosta 33, 66, 115
houseleek 22, 23, 32, 98
Hydrangea anomala 44

I

Iberis sempervirens **137**
ice plant **75**
    see also sedum
Ilex 25, **36**, 116, 120, **137**
insect hotels 105, **124–25**
intercropping 56
iris 33, 46, 119
ivy 9, 24, 25, 107, 114, 120, **136**

J

Japanese barberry 25, **134**
Japanese crab apple **36**
Japanese quince 44, **135**
jasmine 109
Judas tree **35**

K

knapweed 74, 112
knotweed 33

L

lady's leek **126**
lamb's ears 98
landscape materials **84–85**
    see also garden furniture
lavender 33, 41, 97, 98, 107, 109, **130**
lawns **60–61**
    wildflower meadow 60, 61, **112–13**
layering 44
leafmould 31, **51**

leopard plant 115
lettuce 9, 55, 56, 57, 66, 115
lilac **37**, 109
Limnanthes douglasii 43, **130**
Lonicera 22, 109, 124, **130**
loosestrife 115
love-in-a-mist 43, **76**
Lunaria annua 43, 110, **137**
lungwort **76**, 107

M

mahonia 20, 21, **76**, 107
Malus domestica 17, **36**, 58
Malus × floribunda **36**
manure, well-rotted 51, 63
marigold 22, **73**, 83, 90
    French 69
marsh marigold 119
meadowsweet 115
medlar **36**
Mexican fleabane 23, 32
mezereon **136**
mice and voles 123
Michaelmas daisy **73**
mint 69, 109
    catmint **76**, 98, 109
Monarda didyma **131**
mountain clematis **74**
mulching
    and no-dig method 49
    and soil water preservation 98
    water use reduction 98
    see also compost choices
Muscari 107, **131**

N

nasturtium 43, 69
Nectaroscordum siculum **131**
nematodes 68
Nepeta **76**, 98, 109
nettle 13, 63, 64, 109
New England aster **139**
new plant purchase **46–47**
    see also planting and placing
New Zealand bur 23
Nigella damascena 43, **76**
night-scented stock 109
no-dig method 49
Nyssa sylvatica **138**

O

oregano 23
Oregon grape see mahonia
organic certification 58
organic insecticides 69
ox-eye daisy 112

P

Papaver rhoeas 43, 112, **131**
particulates 24, 97
paving 23, 84
pear tree 37
peas see beans and peas
perennials
    division 44
    weed control 65
    woodland, mini 21
pest control **66–69**
    aphids 66, 68, 69, 71, 123
    barrier methods 69
    beer traps 69
    biological controls 68
    and biosecurity 47
    and birds 66, 68
    companion plants 69
    earwigs 69
    and frogs and toads 66, 67
    grit or coffee grounds 69
    inverted pots 69
    nematodes 68
    organic insecticides 69
    pond pests 67
    predator encouragement 66, 105
    slugs and snails 66, 67, 68, 69, 123
    soap 69
    vine weevils 67, 68
    see also planting and placing
pest- and disease-resistant plants 72–77
    allium 21, 43, **72**, 109
    bergenia 21, **73**
    bloody cranesbill 75
    catmint **76**, 98, 109
    Chinese anemone **72**
    clematis 44, **74**, 95, **135**
    cornflower **74**, 112
    ferns 21, 22, 66, **74**
    fuchsia 21, **74**
    hellebore 21, **75**
    love-in-a-mist 43, **76**
    lungwort **76**, 107
    mahonia 20, 21, **76**, 107
    marigold 22, **73**, 83, 90
    Michaelmas daisy **73**
    pincushion flower **77**
    rose, rambling **77**
    rosemary **77**, 97
    sedum 22, 33, 66, **75**, 98, 107
    snapdragon **72**, 107
    sweet woodruff 21, **75**
    trailing bellflower 23, **73**
    zinnia **77**, 83, 107
pesticides
    and bees 106
    organic 69
    and wildlife 13

petunia 108, 109
Phacelia tanacetifolia **132**
Phlomis russeliana **132**
photosynthesis 16
pickerel weed 119
pincushion flower **77**
plant support 95
planters, upcycled **88–89**
planting and placing **40–41**
    soil acidity check 41
    soil types 40
    trees see trees, choosing and planting
    vegetables see vegetable growing
    see also new plant purchase; pest control
planting preparation **48–49**
    compost see compost choices
    new beds 48
    no-dig method 49
    weed removal 48
    worms, encouragement of 49
plastics, reducing and reusing 11, **80–81**
    bioplastics 81
    homemade pots see pots, homemade
    plant-pot recycling schemes 46, 81
    repurposing 80
poached egg plant 43, **130**
pollinator plants
    Balkan clary **133**
    bergamot **131**
    butterfly bush 98, 109, 124, **127**
    butterfly flower **126**
    coneflower see coneflower
    Cornelian cherry **127**
    cosmos 67, 83, 107, **128**
    cowslip **132**
    cranesbill see cranesbill
    crocus 107, **128**
    cuckoo flower **127**
    dahlia 66, 69, 83, **128**
    fiddleneck **132**
    flowering currant 24, 25, **132**
    foxglove 21, **128**
    grape hyacinth 107, **131**
    heather **127**
    hollyhock 42, 43, **126**
    honeysuckle 22, 109, 124, **130**
    lady's leek **126**
    lavender 33, 41, 97, 98, 107, 109, **130**
    poached egg plant 43, **130**
    poppy 43, 112, **131**
    purple top 124, **133**
    red campion 21, **133**
    Sicilian honey garlic **131**

pollinator plants *cont.*
  sneezeweed **130**
  Turkish sage **132**
  winter aconite 107, **129**
pollution traps **24–25**
  air pollution 24
  hedge planting 25
  particulates 24, 97
ponds see garden ponds
poppy 43, 112, **131**
pot marigold 22, 69, **73**, 83, 90
potato and tomato blight 71
pots, homemade 81, **82–83**
  plant-pot recycling schemes 46, 81
  see also plastics, reducing and reusing
powdery mildew 71
predator encouragement 66, 105
primrose 21, 115
Primula japonica 115
Primula veris **132**
propagation methods **44–45**
  cuttings 45
  division 44
  layering 44
Prunus avium **138**
Prunus incisa **37**
Pulmonaria **76**, 107
purple top 124, **133**
Pyracantha 25, 110, **138**

Q
quince 44, **135**

R
rainwater recycling 11, 100
raised beds **90–91**
recycling see garden furniture; landscape materials; planters, upcycled; plastics, reducing and reusing; pots, homemade; raised beds
red campion 21, **133**
red cedar 25
Ribes 24, 25, **132**
river birch 25, **34**
rock rose 41, 98
roof planting 23
rose, rambling **77**
rose, red-leaved **138**
rosemary **77**, 97
rowan **37**, 110
Rubus fruticosus 64, **139**
Rudbeckia 9, 33, **133**
rush 119
Russian sage 98
rust 71

S
Salvia nemorosa **133**
Salvia rosmarinus **77**, 97
Scabiosa caucasica **77**
sea holly 98, 109
sea thrift 22, 23
seaweed 63
sedge 33
sedum 22, 33, 66, **75**, 98, 107
seeds, harvesting **42–43**
  germination tips 43
  seed and plant pots 80
  sowing seeds 43
  storage 42
Siberian squill 21
Sicilian honey garlic **131**
Silene dioica 21, **133**
silver birch 13, 25, **134**
Skimmia japonica 33
slugs and snails 66, 67, 68, 69, 123
small gardens, trees see trees for smaller gardens
smooth serviceberry **34**
snapdragon **72**, 107
sneezeweed **130**
snowdrop 20, 21, 107
soap, and pest control 69
soil acidity check 41
soil types 40
solar-powered units 97, 117
Sorbus aucuparia **37**, 110
spindle tree 21, 25, 33, **35**
succulents 9, 22, 88
sunflower 43, 69, 110, **137**
sweet box 20, 21
sweet pea 82, 95
sweet woodruff 21, **75**
Swiss chard 55, 66
Symphyotrichum **139**
Syringa vulgaris **37**, 109

T
thyme 23, 32, 66, 98, 115
tickseed 110, **135**
tobacco plant 109
trailing bellflower 23, **73**
trees, choosing and planting **18–19**
  and DIY furniture 87
  siting 18
  staking 19
  timing 19
  and wildlife 13, 105
  woodland, mini **20–21**
trees and climate **16–17**
  carbon capture 16, 17
  flood defences 31
  global warming 8
  photosynthesis 16

trees for smaller gardens 34–37
  apple tree 17, **36**, 58
  field maple **34**
  Fuji cherry **37**
  hawthorn 16, 24, 25, 35, 120, **135**
  hazel **35**, 44, 107
  holly 25, **36**, 116, 120, **137**
  Japanese crab apple **36**
  Judas tree **35**
  lilac **37**, 109
  medlar **36**
  pear tree **37**
  river birch 25, **34**
  rowan **37**, 110
  smooth serviceberry **34**
  spindle tree 21, 25, 33, **35**
tupelo **138**
Turkish sage **132**

U
understorey planting 20
upcycling see *under* recycling

V
vegetable growing 9, **54–57**
  beans and peas see beans and peas
  cabbage family 57, 69, 99
  companion plants 69
  containers 55
  crop choice 56
  crop rotation 57, 70
  green manure, planting 55
  intercropping 56
  pests 68, 69, 90
  plot location 54
  plot preparation 55
  raised beds, recycled **90–91**
  root vegetables 57, 101
  simple crop plan 57
  top-dressing and compost 55
  weeding 55
Verbena bonariensis 124, **133**
vertical gardens 22
viburnum 20, 21, 25, 110, 114
vine weevils 67, 68
viola 33

W
water avens 115
water butt 11, 95, 100
water lily 28, 118
water use reduction **98–101**
  and climate conditions 98
  grey water 101
  mulching 98

rainwater 11, 100
  storage 11, 95, 100, 101
  watering frequency 99
water for wildlife **116–17**
  see also wildlife, feeding
weed control **64–65**
  composting 65
  ground cover 64
  lawns 60
  perennials 65
  planting preparation 48
  vegetable growing 55
  and wildlife 13
weigela 21
wild cherry **138**
wild garlic 20, 21
wild marjoram 115
wildflower meadow 60, 61, **112–13**
wildlife, designing for **12–13**, 20
wildlife, feeding
  bees see bees
  birds see birds
  bog garden 115
  butterflies and moths 107, **108–09**, 114
  frogs, toads and newts 123
  garden ecosystem 104, 114
  garden predators 66, 105
  habitat increase **114–15**
  insect hotels 105, **124–25**
  mice and voles 123
  pollinator see pollinator plants
  ponds see garden ponds
  small creatures **122–23**
  and trees 13, 105
  water for wildlife **116–17**
  wildflower meadow 60, 61, **112–13**
  woodlice and beetles 122, 123
  woodpiles 123, 124
  see also birds and wildlife plants
winter aconite 107, **129**
wisteria 22
wood accreditation 85
wood anemone 21
wood spurge 21, 33
woodland, mini **20–21**
  see also trees headings
woodlice and beetles 122, 123
woodpiles 123, 124
worms 49

Y
yarrow **134**
yew 24, 25, 28, 114, 120, **139**

Z
zinnia **77**, 83, 107

BIBLIOGRAPHY

page 8–9 Sánchez-Bayo, F. and Wyckhuys, Kris A.G., 'Worldwide decline of the entomofauna: A review of its drivers', *Biological Conservation* (April 2019) 232: 8-27

'Outdoor Air and Pesticides', National Pesticide Information Centre, www.npic.orst.edu/envir/outair.html

pages 16–17 'Heat Island Impacts', United States Environment Protection Agency, www.epa.gov/heatislands/heat-island-impacts

pages 26–27 Jeffries, M., 'Ponds can absorb more carbon than woodland', Northumbria University, www.northumbria.ac.uk/about-us/news-events/news/ponds-absorb-carbon

Finlay, K. and McCowan, C., 'Farm ponds can act as greenhouse gas sinks in the Canadian Prairies', www.theconversation.com/farm-ponds-can-act-as-greenhouse-gas-sinks-in-the-canadian-prairies-115058

Peacock, M. et al., 'Greenhouse gas emissions from urban ponds are driven by nutrient status and hydrology', *Ecosphere*, https://esajournals.onlinelibrary.wiley.com/doi/full/10.1002/ecs2.2643

pages 48–49 'Earthworms', Garden Organic, www.gardenorganic.org.uk/earthworms

Turetsky, M. R. et al., 'Global vulnerability of peatlands to fire and carbon loss', *Nature Geoscience* (2015), 8: 11–14

pages 60–61 Son, J., 'Lawn maintenance and climate change', Princeton Student Climate Initiative, https://psci.princeton.edu/tips/2020/5/11/law-maintenance-and-climate-change

pages 80–81 'How does plastic harm the environment?' Friends of the Earth, https://friendsoftheearth.uk/plastics

'Plastics in the Garden', Garden Organic, www.gardenorganic.org.uk/plastics-garden

pages 84–85 Watts, J., 'Concrete: the most destructive material on Earth', *The Guardian*, 25 Feb 2019, www.theguardian.com/cities/2019/feb/25/concrete-the-most-destructive-material-on-earth

Lehne, J. and Preston, F., 'Making Concrete Change; Innovation in Low-carbon Cement and Concrete', Chatham House Report (2018) www.chathamhouse.org/sites/default/files/publications/research/2018-06-13-making-concrete-change-cement-lehne-preston.pdf

pages 96–97 Hawkins, J. L. et al., 'Exercise Intensities of Gardening Tasks Within Older Adult Allotment Gardeners in Wales', *Journal of Aging and Physical Activity*, 23: Issue 2, 161–168

Harrabin, R. 'Wood burners: Most polluting fuels to be banned In the home', BBC, 21 February 2020, www.bbc.co.uk/news/uk-51581817

pages 100–101 Belanger, J. D., 'Recycling Water at Home: Grey Water for the Garden' *Countryside Magazine* (February 25, 2019) https://iamcountryside.com/growing/recycling-water-grey-water-for-the-garden

pages 104–105 Thompson, K., Wildlife Gardening Forum: 'The complexity of garden food webs', www.wlgf.org/food_webs.html

'Biodiversity in Urban Gardens in Sheffield', University of Sheffield, www.bugs.group.shef.ac.uk/BUGS1/bugs1-index.html

pages 106–107 'Facts About Bees', The Soil Association, www.soilassociation.org/organic-living/bee-organic/10-facts-about-bees

'The NAPPC Honey Bee Health Task Force', Pollinator Partnership, www.pollinator.org

Centre for Biological Diversity, www.biologicaldiversity.org

pages 110–111 'Big Butterfly Count 2020: The Results', Butterfly Conservation, www.butterfly-conservation.org/news-and-blog/big-butterfly-count-2020-the-results

'Butterfly declines in North America and the UK', www.dw.com/en/butterfly-declines-in-the-north-america-uk/a-37688619

pages 114–115 'Biodiversity and Ecosystems', Stand For Trees, www.standfortrees.org/why-it-matters/biodiversity-ecosystems

pages 116–117 Head, S. and Thomas, A., 'Introduction to water in the garden and pond ecology', Wildlife Gardening Forum, www.wlgf.org/water_ecology.html

RESOURCES

If you would like more information on some of the topics in this book, the following organizations can help.

For general information about ecological issues, eco gardening, and healthy soils:

Charles Dowding (no-dig methods)
https://charlesdowding.co.uk

Friends of the Earth
https://friendsoftheearth.uk

Garden Organic
www.gardenorganic.org.uk

Soil Association
www.soilassociation.org

Sustainability Guide
www.sustainabilityguide.co.uk

The Woodland Trust
www.woodlandtrust.org.uk

For more on plant and animal conservation in your garden:

Amphibian and Reptile Conservation Trust
www.arc-trust.org

Bee Friendly Trust
www.beefriendlytrust.org

British Trust for Ornithology
www.bto.org

Bumblebee Conservation Trust
www.bumblebeeconservation.org

Butterfly Conservation
www.butterfly-conservation.org

Freshwater Habitats Trust
www.freshwaterhabitats.org.uk

Plantlife
www.plantlife.org.uk

Repollinate
www.repollinate.org.uk

Rewilding Britain
www.rewildingbritain.org.uk

Royal Society for the Protection of Birds
www.rspb.org.uk

UK Moths Count
www.ukmoths.org.uk

The Wildlife Trusts
www.wildlifetrusts.org

Wildlife Gardening Forum
www.wlgf.org

Author Zia Allaway

AUTHOR ACKNOWLEDGMENTS

Many thanks to Marek Walisiewicz at cobalt id for commissioning me to write this inspirational book and to Paul Reid for his beautiful designs. I would also like to thank editor Diana Vowles for her skill, patience and good humour, and Amy Slack at Dorling Kindersley for her attention to detail.

PUBLISHER ACKNOWLEDGMENTS

DK would like to thank Oreolu Grillo and Sophie State for early spread development for the series, and Margaret McCormack for indexing.

PICTURE CREDITS

The publisher would like to thank the following for their kind permission to reproduce their photographs:

Alamy Stock Photo: Anton Garin 4c; blickwinkel 6c; JosephWGallagher 8tl; Jean Williamson 10br; Miriam Heppell 11tl; Derek Harris 13cl; Annie Eagle 13br; kris Mercer 26bl; Tim Gainey 35tl; Christina Bollen 36bl; GKSFlorapics 51tr; Pavol Klimek 61cr; john t. fowler 67br; Tim Gainey 81cr; Tim Gainey 82tr; David Burton 84cr; Avalon/Photoshot License 85bc; Arcaid Images 86bl; Sawangwit Muanghtai 86br; Panther Media GmbH 87bl; Pixelot 88tr; Heather Edwards 92cl; Zoonar GmbH 93tr; keith burdett 94tr; Goddard New Era 95tr; Design Pics Inc 96bl; Francisco Martinez 97cl; Jurate Buiviene 102c; SelectPhoto 106cl; David Chapman 110br; Anna Stowe Botanica 111tl; Theo Moye 111bc; Sue Robinson 116tr; Gillian Pullinger 116cl; idp wildlife collection 116br; David Stuckel 120cl; gary corbett 121bl; Yon Marsh Natural History 122cl.

Dorling Kindersley: 123RF.com / Gunnar Pippel / gunnar3000 112br; 123RF.com / Leonid Ikan 108tr; 123RF.com / Pumidol Leelerdsakulvong 30cl; Alan Buckingham 71br; Brian North 89br, 99cl; Brian North / RHS Chelsea Flower Show 12tr, 120tr; Brian North / RHS Hampton Court Flower Show 8br, 32bl, 52cr, 84tr, 87tr, 88bc, 133br, 101tr, 86tr; Brian North / Waterperry Gardens 24bl; Debbie Patterson / Ian Cuppleditch 25tl, 135br; Dreamstime.com / Dave Massey / Dmass 106tr; Dreamstime.com / Dfikar 108br; Dreamstime.com / Jochenschneider 26br; Dreamstime.com / Ker784 65t; Dreamstime.com / Richard J Thompson / Photoaged 108tl; Dreamstime.com / Valentino2 123tc; iStock / Pavliha 17tr; Jerry Harpur / National Trust (Erdigg) 136br; Kim Taylor 110cl; Mark Winwood 115cl, 115tl, 115tr; Mark Winwood / Ball Colegrave 72br, 77br, 128tl; Mark Winwood / Downderry Nursery 130tr; Mark Winwood / Dr Mackenzie 21tr; Mark Winwood / Hadlow College 75bl; Mark Winwood / John Hall Plants, Hindhead 127tr, Mark Winwood / Marle Place Gardens and Gallery, Brenchley, Kent 127br; Mark Winwood / RHS Chelsea Flower Show 127tl, 129tl; Mark Winwood / RHS Malvern Flower Show 74tl, Mark Winwood / RHS Wisley 9br, 18tr, 34tr, 36tl, 37bl, 72tr, 74br, 75br, 76br, 76tl, 76tr, 107tc, 107tr, 109bc, 109tr, 114tr, 130tl, 133bl, 133tl, 137bl, 138tr; Peter Anderson 11bl, 20bl, 20tr, 24cr, 32bc, 35tr, 63tl, 84bl, 85tr, 115cr, 122br, 134br; Peter Anderson / National Dahlia Collection 128bl; Peter Anderson / RHS Chelsea Flower Show 22br; Peter Anderson / RHS Hampton Court Flower Show 22tc, 31tr, 33br, 49tr, 85cl, 98bc, 105br, 105tl, 112tr, 124bl; RHS Tatton Park 40tr, 131tl, 135bl; RHS Wisley 35br; Steve Hamilton / Chelsea Physic Garden, London 136tl.

GAP Photos: Marcus Harpur - Design: Wendy Allen Hadlow College with Westgate Joinery 32tr.

Getty Images: sassy1902 30br; xavierarnau 46cl; welcomia 47bl; Natthapong Daeng Leis / EyeEm 100tl.

Cover images: Back: Dreamstime.com: Richard J Thompson / Photoaged cl

Illustrations by cobalt id.

All other images © Dorling Kindersley

Produced for DK by
COBALT ID
www.cobaltid.co.uk

Managing Editor Marek Walisiewicz
Editor Diana Vowles
Managing Art Editor Paul Reid
Art Editor Darren Bland

DK LONDON
Project Editor Amy Slack
Managing Editor Ruth O'Rourke
Managing Art Editor Christine Keilty
Production Editor David Almond
Production Controller Stephanie McConnell
Jacket Designer Nicola Powling
Jacket Co-ordinator Lucy Philpott
Art Director Maxine Pedliham
Publishers Mary-Clare Jerram, Katie Cowan

First published in Great Britain in 2021 by
Dorling Kindersley Limited
DK, One Embassy Gardens, 8 Viaduct Gardens,
London, SW11 7BW

A CIP catalogue record for this book
is available from the British Library.
ISBN: 978-0-2414-5861-7

Printed and bound in China

For the curious
www.dk.com